AS-Level Biology AQA B

The Revision Guide

Editors:
Katherine Reed, Kate Houghton

Contributors:
Gloria Barnett, Claire Charlton, Martin Chester, Barbara Green, Anna-Fe Guy, Dominic Hall, Gemma Hallam, Becky May, Stephen Phillips, Kate Redmond, Claire Reed, Adrian Schmit, Rachel Selway, Emma Singleton, Sharon Watson.

Proofreaders:
Ben Aldiss, Vanessa Aris, James Foster, Tom Trust.

Published by Coordination Group Publications Ltd.

ISBN: 1 84146 954 8

Groovy website: www.cgpbooks.co.uk
Jolly bits of clipart from CorelDRAW
Printed by Elanders Hindson, Newcastle upon Tyne.

Contents

Carbohydrates

Hello. Welcome to your revision guide for this evening. Get yourself a glass of water and open the window. Then settle down to learn about amazing, fascinating... carbohydrates. All carbohydrates contain only carbon, hydrogen and oxygen. They're needed for energy and some, like cellulose, give structural support. Life doesn't get any better than this...

Carbohydrates are Made from **Monosaccharides**

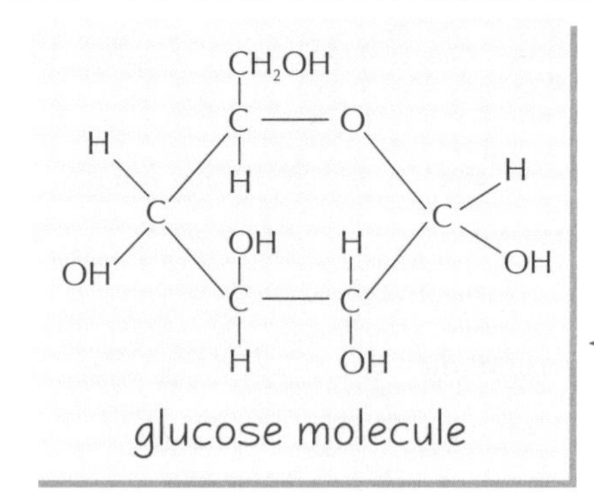

All carbohydrates are made from sugar molecules. A single sugar molecule is called a **monosaccharide**. Examples of monosaccharides include glucose, fructose, ribose, deoxyribose and galactose. These are the basic molecular units (called **monomers**) which other carbohydrates are made up of. The only monosaccharide you need to learn the structural formula for is glucose, which isn't too bad really.

Disaccharides are **Two Monosaccharides** Joined Together

Disaccharides are sugars made from two monosaccharide sugar molecules stuck together. Examples are:

DISACCHARIDE	MONOSACCHARIDES IT'S MADE UP OF
maltose	glucose + glucose
sucrose	glucose + fructose

Extensive scientific research revealed an irreversible bond joining sugars to Pollyanna's gob.

Condensation Reactions Join Sugars Together

When the sugars join, a molecule of water is squeezed out. This is called a **condensation reaction**. The bonds that join sugars together are called **glycosidic bonds**.

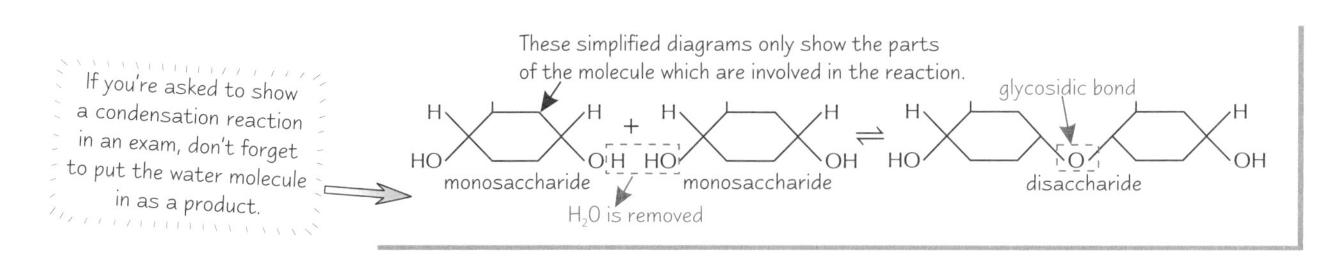

Hydrolysis Breaks Sugars Apart

When sugars are separated, the condensation reaction goes into **reverse**. This is called a **hydrolysis reaction** — a water molecule reacts with the glycosidic bond and breaks it apart.

disaccharide

monosaccharide monosaccharide

H_2O reacts with glycosidic bond

It's all in the name — "hydro" is to do with water, and "lysis" means breaking up.

Condensation and hydrolysis reactions are dead important in biology. **Proteins and lipids** are put together and broken up by them as well. So you definitely need to understand how they work.

Carbohydrates

Polysaccharides are **Loads of Sugars** Joined Together

Polysaccharides are molecules which are made up of **loads of sugar molecules** stuck together.
The ones you need to know about are:

1) **starch** — the main storage material in plants;
2) **glycogen** — the main storage material in animals;
3) **cellulose** — the major component of cell walls in plants.

Examiners like to ask about the link between the structures of polysaccharides and their functions.

(1) **Starch** is made up of **two** other polysaccharides of **glucose**:

- **Amylose** is a long, **unbranched chain** of glucose. The angles of the glycosidic bonds give it a **coiled structure**, almost like a cylinder. Its **compact**, coiled structure makes it really **good for storage**.
- **Amylopectin** is a long, **branched chain** of glucose. Its **side branches** make it particularly good for the storage of glucose — the enzymes that break down the molecule can get at the glycosidic bonds easily, to break them and release the glucose.
- Starch is a **large** molecule which is **insoluble** in water. This means that it can be stored in large amounts without really affecting the water potential of cells (see p.19) — so it's good for storage.

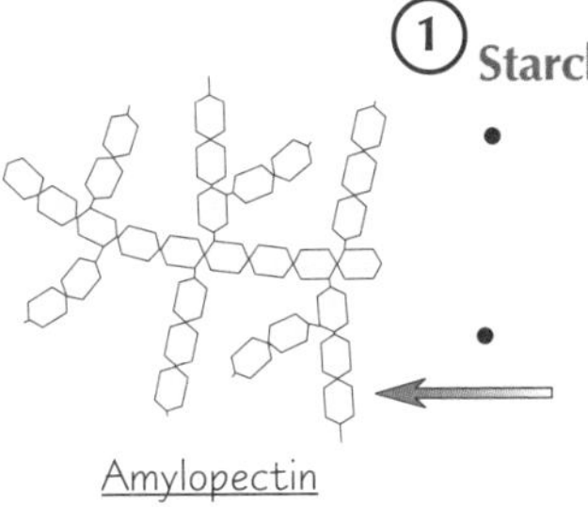
Amylopectin

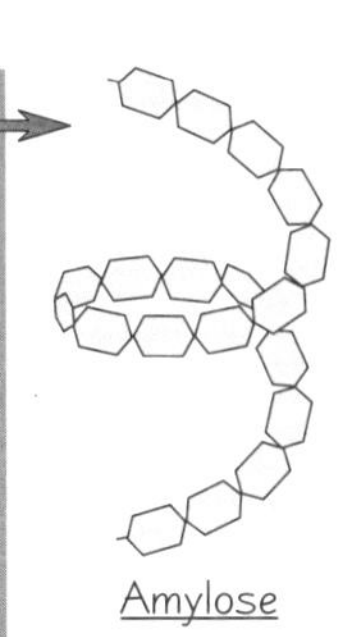
Amylose

(2) **Glycogen** is a polysaccharide of **glucose**. Its structure is very similar to amylopectin, except that it has **loads** more **side branches** coming off it. It's a very **compact** molecule found in animal liver and muscle cells so it's good for storage. Loads of branches mean that stored glucose can be released quickly, which is **important for energy release** in animals.

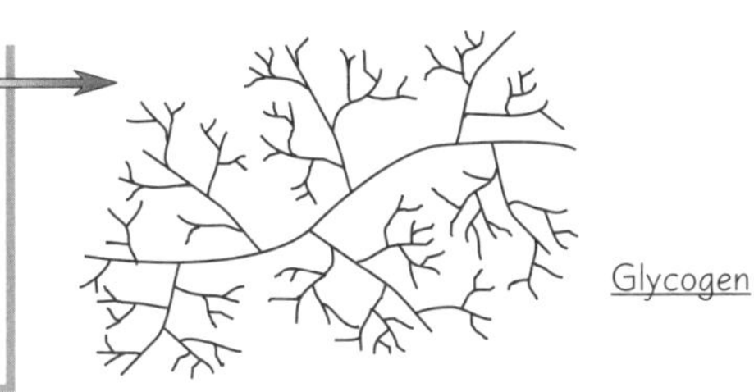
Glycogen

3 Cellulose molecules

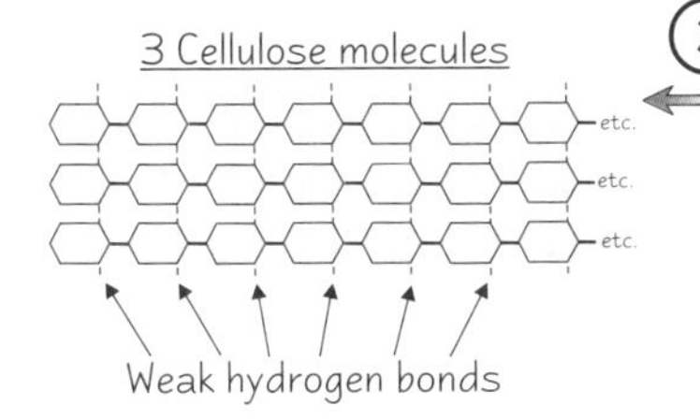

(3) **Cellulose** is made of long, unbranched chains of **glucose**. The bonds between the sugars are **straight**, so the chains are straight. The chains are linked together by **hydrogen bonds** (see p.10) to form strong fibres called **microfibrils**. The strong fibres mean cellulose can provide **structural support** for cells. Another feature is that the **enzymes** that break the glycosidic bonds in starch can't reach the glycosidic bonds in cellulose, so those enzymes **can't break down cellulose**.

Practice Questions

Q1 What is the name given to the type of bond that holds sugar molecules together?

Q2 Name the two different types of molecule that are combined together in a starch molecule.

Q3 Name three polysaccharides and give the function of each one.

Q4 How does cellulose's structure make it good at providing structural support in plant cells?

Exam Questions

Q1 Describe how glycosidic bonds in carbohydrates are formed and broken in living organisms. [7 marks]

Q2 Compare and contrast the structures of glycogen and cellulose,
showing how each molecule's structure is linked to its function. [10 marks]

Who's a pretty polysaccharide, then...

If you learn these basics it makes it easier to learn some of the more complicated stuff later on — 'cos carbohydrates crop up all over the place in biology. Remember that condensation and hydrolysis reactions are the reverse of each other — and don't forget that starch is composed of two polysaccharides. So many reminders, so little space ...

Proteins

There are hundreds of different proteins — all of them contain carbon, hydrogen, oxygen and nitrogen. They are the most abundant organic molecules in cells, making up 50% or more of a cell's dry mass — now that's just plain greedy.

Proteins are Made from Long Chains of Amino Acids

1) The **monomers** (basic molecular units) of proteins are **amino acids**. There are about **20** different amino acids in humans.
2) **Two amino acids** joined together form a **dipeptide**. **Loads of amino acids** joined together form a **polypeptide**. A **protein** is made up of one or more polypeptides.
3) All amino acids have a **carboxyl group** (-COOH) and an **amino group** ($-NH_2$) attached to a carbon atom.

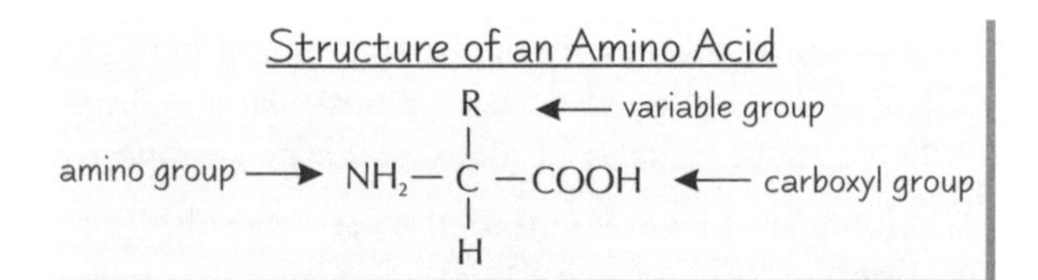

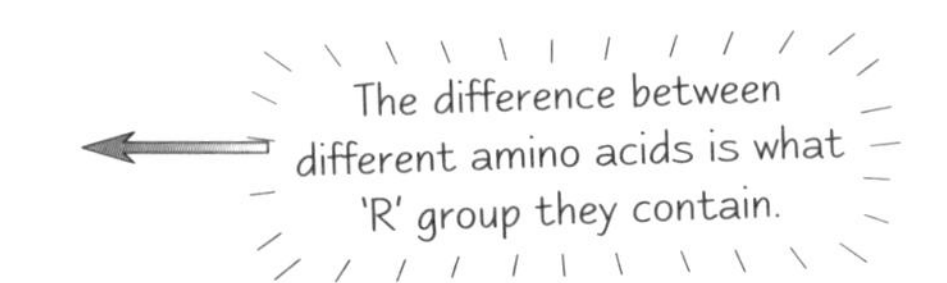

Proteins are Formed by Condensation Reactions

Just like carbohydrates, the parts of a protein are put together by **condensation** reactions and broken apart by **hydrolysis** reactions. The bonds that are formed between amino acids are called **peptide bonds**.

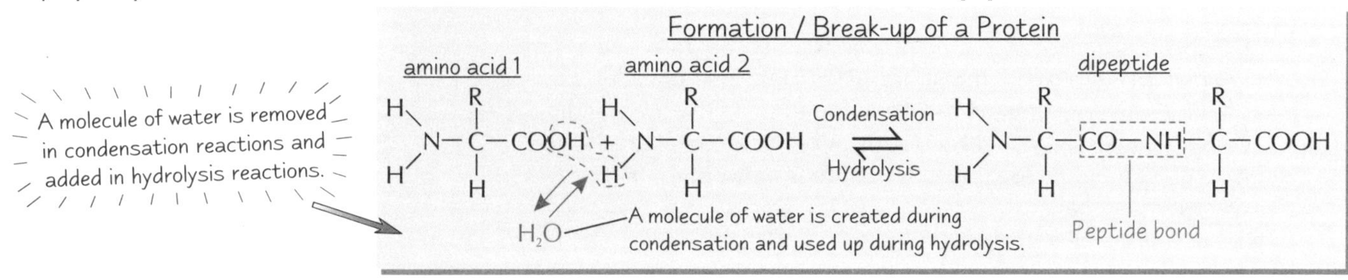

Proteins have a Primary, Secondary and Tertiary Structure

Proteins are **big**, **complicated** molecules. They're easier to explain if you describe their structure in three 'levels'. These levels are called the protein's **primary, secondary** and **tertiary** structures.

1. The **primary structure** is the **sequence of the amino acids** in the long chain that makes up the protein (the **polypeptide chain**).

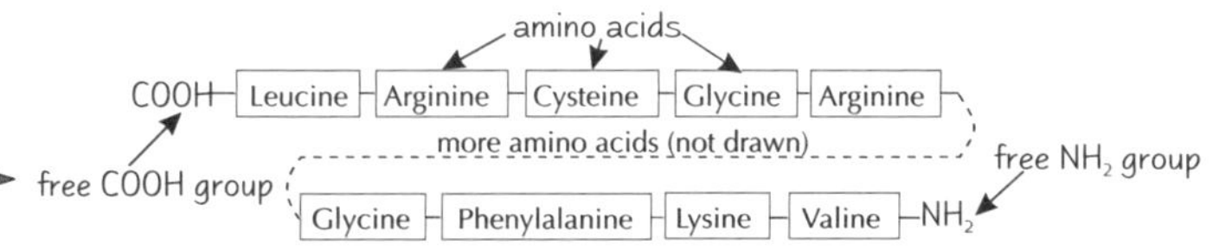

2. The **bonds** between the amino acids make the chain form a sort of **coil**. The way the chain coils is called its **secondary structure**. The most common secondary structure is a **spiral** called an **alpha (α) helix**.

3. The coiled chain of amino acids is itself often coiled and folded in a characteristic way that identifies the protein. **Extra bonds** can form between different parts of the polypeptide chain, e.g. hydrogen bonds, ionic bonds and disulphide bonds. These give the protein a kind of **three dimensional shape**. This is its **tertiary structure**.

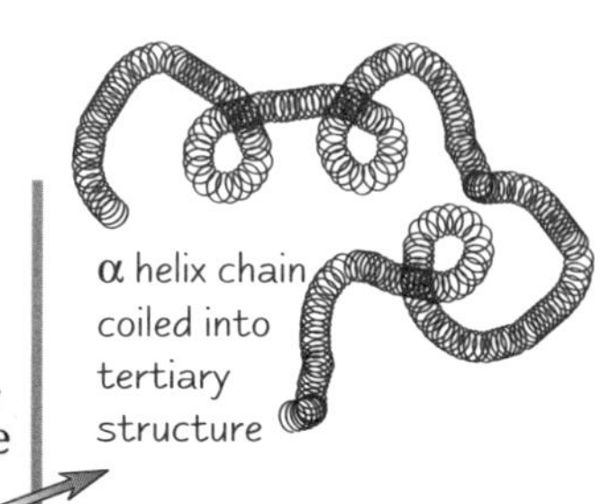

Proteins

Protein Shape Relates to its **Function**

There are two kinds of protein:

1) **Globular proteins** are round and compact. They have a complicated **tertiary structure** with **lots of folds** in the polypeptide chain. They're **soluble** so they are easily transported around by the blood.
2) **Fibrous proteins** consist of **long, polypeptide chains**. The chains form a **simple tertiary structure** where the polypeptide chains coil round each other to form **strong, insoluble fibres**.

You need to learn a few **examples** of how proteins are **adapted for their jobs**.

Collagen is a **fibrous protein** that forms **supportive tissue** in animals, so it needs to be strong.

Haemoglobin is a **globular protein** that absorbs oxygen. Its structure is curled up, so **hydrophilic** ('water-attracting') side-chains are on the **outside** of the molecule and **hydrophobic** ('water-repelling') side-chains face **inwards**. This makes them soluble in water and good for transport in the blood.

Insulin is a hormone that reduces blood glucose levels.

Proteins can be **Denatured** by Temperature or pH Changes

1) The structure of a protein can be damaged by **temperature** or **pH changes**. This is called **denaturation**.
2) This **damage** means that the protein can no longer do its job, e.g. enzymes aren't able to catalyse reactions after being denatured.

Practice Questions

Q1 Name the four elements that are found in all proteins.

Q2 What is a dipeptide?

Q3 Draw the chemical structure of an amino acid.

Q4 Why are globular proteins easy to transport in the blood?

Exam Questions

Q1 Describe the structure of a protein, explaining the terms primary, secondary and tertiary structure. [7 marks]

Q2 Describe the structure of the collagen molecule, and explain how this structure relates to its function in the body. [6 marks]

"It ain't the Taj Mahal or the Hanging Gardens of Babylon, but it'll do..."

What part of Jane Russell was Robert Mitchum referring to, in this immortal line from the Joseph Von Sternberg/Nicholos Ray directed 'Macao'? Answers on a postcard please. Meanwhile... do you know what a monomer and a polymer are? Monomer = basic molecular unit (of a protein, carbohydrate or lipid). Polymer = lots of monomers joined together.

Lipids

Lipids are fats and oils — they are all made up of carbon, hydrogen and oxygen, and they're all insoluble in water. Ever seen a lump of butter dissolve in water? No — exactly.

Lipids *are Fats and Oils — they're* ***Useful***

1) Lipids contain a lot of **energy per gram**, so they make useful **medium** or **long-term energy stores**. But they can't be broken down very quickly, so organisms use carbohydrates for **short-term storage**.
2) Lipids stored under the skin in **mammals** act as **insulation**. Skin loses heat from blood vessels, but the fatty tissue under the skin doesn't have an extensive blood supply, so it conserves heat.

Most Fats and Oils are ***Triglycerides***

Most lipids are composed of compounds called **triglycerides**. Triglycerides are composed of one molecule of **glycerol** with **three fatty acids** attached to it.

Structure of a Triglyceride

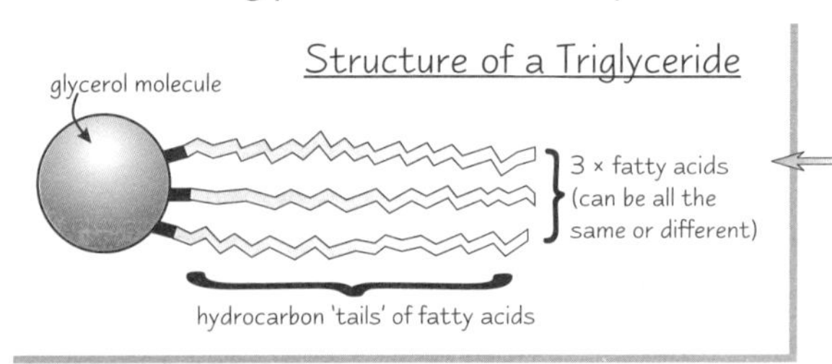

Fatty acid molecules have long 'tails' made of **hydrocarbons**. The tails are '**hydrophobic**' (they repel water molecules). These tails make lipids insoluble in water. When put in water, fat and oil molecules **clump together** in globules to reduce the surface area in contact with water.

All **fatty acids** consist of the same basic structure, but the **hydrocarbon tails vary.** The tail is shown in the diagram with the letter 'R'.

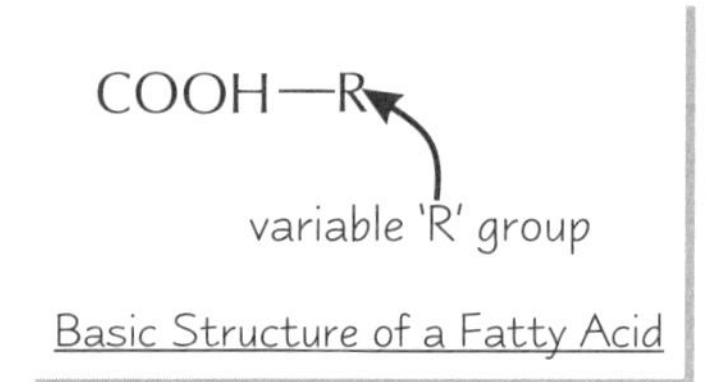

Basic Structure of a Fatty Acid

Triglycerides are ***Formed*** *by... you guessed it...* ***Condensation Reactions***

Like carbohydrates and proteins, lipids are formed by **condensation reactions** and broken up by **hydrolysis reactions**.

1) The diagram below shows a **fatty acid** joining to a **glycerol molecule**.
2) The **bond** formed between glycerol and a fatty acid is called an **ester bond**.
3) A molecule of water is also formed — it's a **condensation reaction**. This process happens twice more, to form a **triglyceride**.
4) The **reverse** happens in **hydrolysis** — a molecule of water is added to each ester bond to break it apart, and the triglyceride splits up into three fatty acids and one glycerol molecule.

Formation of a Triglyceride

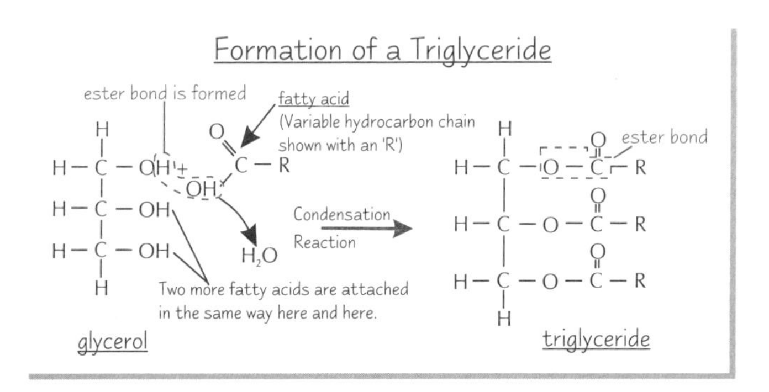

Lipids

Lipids can be **Saturated** or **Unsaturated**

There are two kinds of lipids — **saturated** lipids and **unsaturated** lipids. **Saturated** lipids are mainly **animal fats** and **unsaturated** lipids are found mostly in **plants** (unsaturated lipids are called **oils**). The difference between these two types of lipids is in the **hydrocarbon tails** of the fatty acids.

1) Saturated fats **don't** have any **double bonds** between their carbon atoms — every bond has a **hydrogen** atom attached. The lipid is 'saturated' with hydrogen.
2) Unsaturated fats **do** have double bonds between carbon atoms. If they have **two or more** of them, the fat is called **polyunsaturated** fat.

Unsaturated fats melt at lower temperatures than saturated ones. When used in margarine or butter spreads, it makes them easier to use 'straight from the fridge'.

Phospholipids are a Special Type of Lipid

The lipids found in **cell membranes** aren't triglycerides — they're **phospholipids**. The difference is small but important:

1) In phospholipids, a **phosphate group** replaces one of the fatty acid molecules.
2) The phosphate group is **ionised**, which makes it **attract water** molecules.
3) So part of the phospholipid molecule is **hydrophilic** (attracts water) while the rest (the fatty acid tails) is **hydrophobic** (repels water). This is important in the cell membrane (see p.16 to find out why).

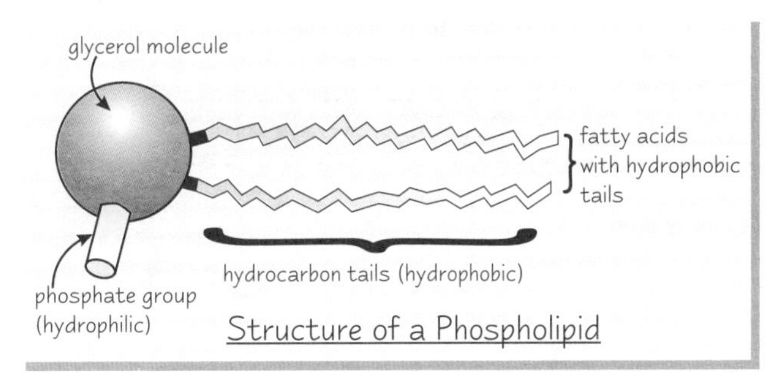

Structure of a Phospholipid

Practice Questions

Q1 Why is it wrong to call lipids "fats"?

Q2 Why are lipids insoluble in water?

Q3 What's the name given to the type of bond that joins fatty acids to glycerol in a lipid molecule?

Q4 Explain the difference between a triglyceride and a phospholipid.

Exam Questions

Q1 Describe the chemical reactions involved in the assembly and break down of triglycerides in living organisms. [8 marks]

Q2 Describe the differences between a triglyceride and a phospholipid, and explain how these differences affect the properties of the molecule. [8 marks]

I eat chips, therefore I am...

You don't get far in life without extensive lard knowledge, so learn all the details on this page good and proper. Lipids pop up in other sections, so make sure you know the basics about how their structure gives them some quite groovy properties. Right, all this lipid talk is making me hungry...

Biochemical Tests for Molecules

Here's a bit of light relief for you — two pages all about how you test for different food groups. There's nothing very complicated, you just need to remember a few chemical names and some colour changes.

Use the **Benedict's Test** for **Sugars**

The Benedict's test identifies **reducing sugars**. These are sugars that can donate electrons to other molecules — they include **all monosaccharides** and **some disaccharides**, e.g. maltose. When added to reducing sugars and heated, the **blue Benedict's reagent** gradually turns **brick red** due to the formation of a **red precipitate**.

The colour changes from:

blue → **green** → **yellow** → **orange** → **brick red**

The higher the concentration of reducing sugar, the further the colour change goes — you can use this to **compare** the amount of reducing sugar in different solutions. A more accurate way of doing this is to **filter** the solution and **weigh the precipitate**.

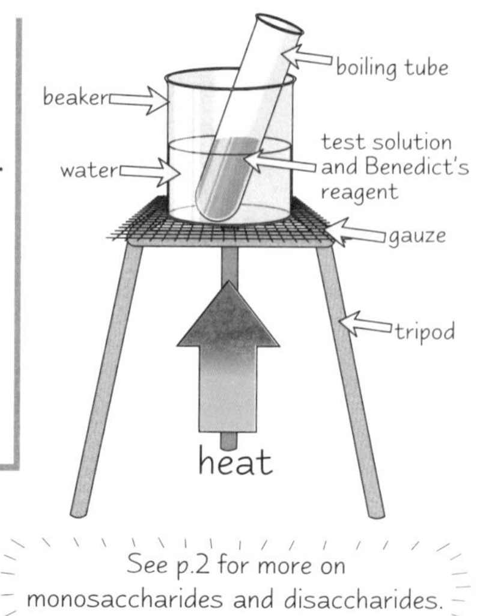

See p.2 for more on monosaccharides and disaccharides.

To test for **non-reducing sugars** like sucrose, which is a disaccharide (two monosaccharides joined together), you first have to break them down chemically into monosaccharides. You do this by boiling the test solution with **dilute hydrochloric acid** and then neutralising it with sodium hydrogen carbonate before doing the Benedict's test.

Use the **Iodine Test** for **Starch**

Make sure you always talk about <u>iodine in potassium iodide solution</u>, not just iodine.

In this test, you don't have to make a **solution** from the substance you want to test — you can use **solids** too. Dead easy — just add **iodine dissolved in potassium iodide solution** to the test sample. If there's starch present, the sample changes from **browny-orange** to a dark, **blue-black** colour.

Use the **Biuret Test** for **Proteins**

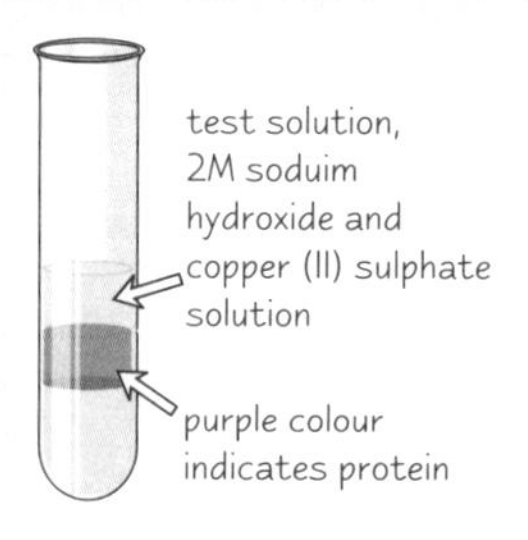

There are **two stages** to this test.

1) The test solution needs to be **alkaline**, so first you add a few drops of **2M sodium hydroxide**.

2) Then you add some **0.5% copper (II) sulphate solution**. If a **purple layer** forms, there's protein in it. If it stays **blue**, there isn't. The colours are pale, so you need to look carefully.

Use the **Emulsion Test** for **Lipids**

Shake the test substance with **ethanol** for about a minute, then pour the solution into water. Any lipid will show up as a **milky emulsion**. The more lipid there is, the more noticeable the milky colour will be.

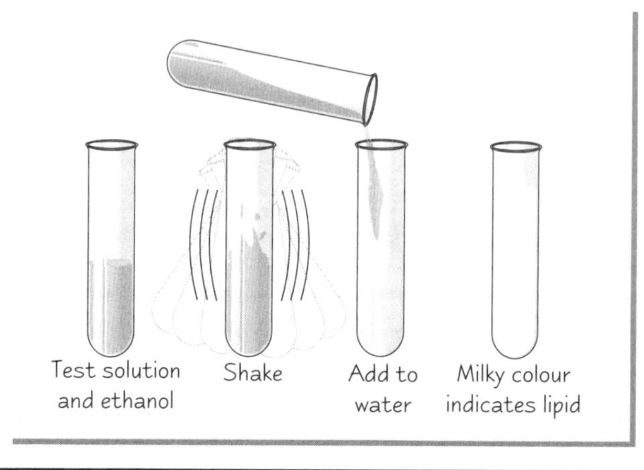

Biochemical Tests for Molecules

Chromatography Separates Out Molecules

If you have a mixture of biological chemicals in a sample that you want to test, you can separate them using the technique of **chromatography**.

1) You put a spot of the test solution onto a strip of special **chromatography paper**, then dip the end of the strip into a **solvent**.
2) As the solvent spreads up the paper, the different chemicals move with it, but at **different rates**, so they separate out.
3) You can identify what the chemicals are using their **Rf values** (see diagram below).

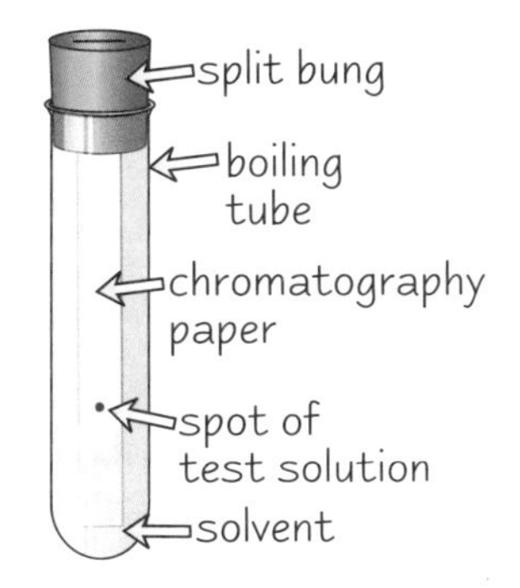

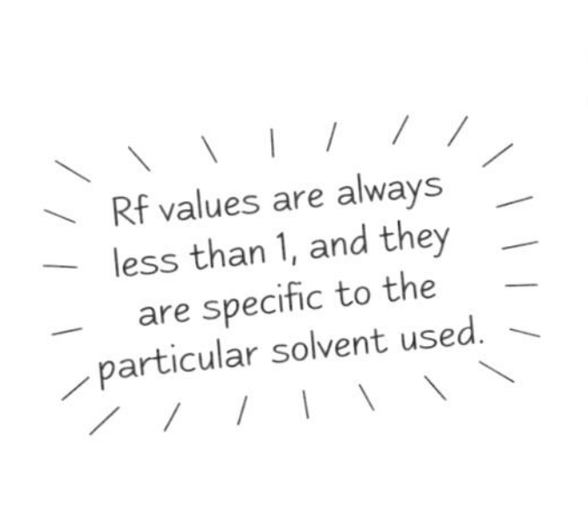

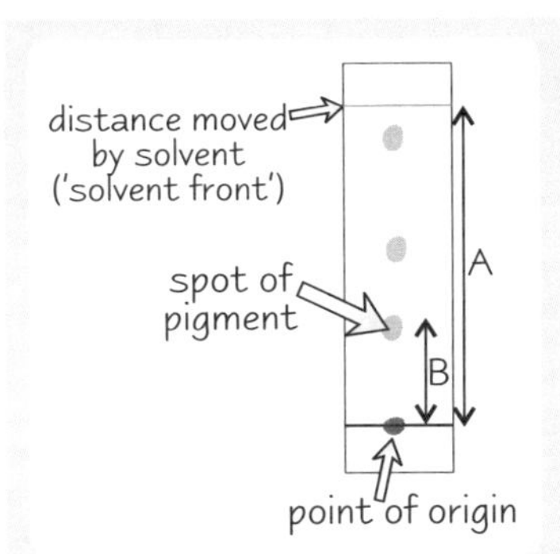

Rf value of pigment = $\frac{B}{A}$

= $\frac{\text{distance travelled by spot}}{\text{distance travelled by solvent}}$

Sometimes, the solvent doesn't completely separate out all the chemicals. In this case you need to use **two way chromatography**, which uses a second solvent to complete the separation.

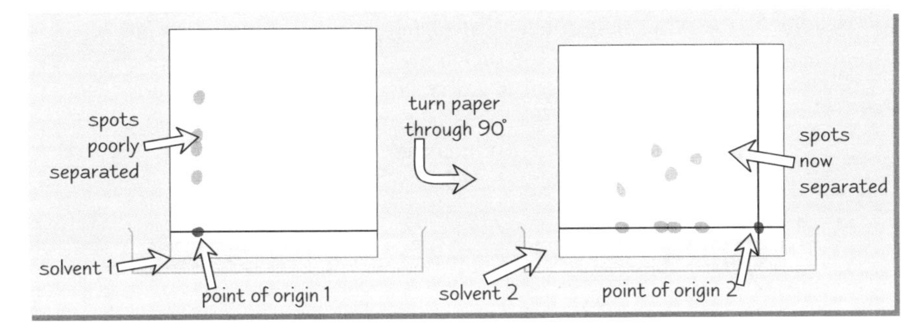

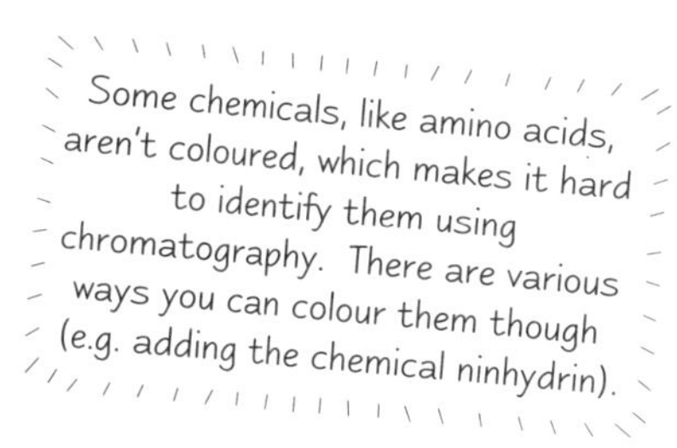

Practice Questions

Q1 How can you work out the different concentrations of reducing sugars in two solutions?

Q2 Describe how you would test a solution for starch. What result would you expect if:
a) starch was present? b) starch was not present?

Q3 How is an 'Rf value' of a chemical calculated?

Q4 When would you use 'two-way' chromatography?

Exam Questions

Q1 You are given an unknown solution to test for different biochemical groups.
Describe the tests you would carry out and how you would analyse the results. [14 marks]

Q2 Describe how you would separate and identify the different pigments in a leaf by means of chromatography. [7 marks]

The Anger Test — Annoy the test subject. If it goes red, anger is present...

The days of GCSEs might have gone forever, but with this page you can almost feel like you're back there in the mists of time, when biology was easy and you fancied someone out of S-Club 7. Aah. Well, you'd better make the most of it and get these tests learnt — 'cos things get trickier than a world-class magician later on...

Water

Life can't exist without water — in fact boring everyday water is one of the most important substances on the planet. Funny old world.

Water is Vital to Living Organisms

Water makes up about 80% of cell contents — it has loads of important **functions**, inside and outside cells.

1) Water is a **metabolic reactant**. That means it's needed for loads of important **chemical reactions**, like photosynthesis and hydrolysis reactions (remember them?).
2) Water is a **solvent** — which means many substances dissolve in it. Most biological reactions take place **in solution**, so water's solvent properties are vital.
3) Water **transports** substances. The fact that it's a **liquid** and a solvent makes it easy for water to transport all sorts of materials around plants and animals, like glucose and oxygen.
4) Water helps with **temperature control**. When water **evaporates**, it uses up heat from the surface that it's on. This cools the surface and helps lower the temperature.

Water Molecules have a Simple Structure

The **structure of a water molecule** helps to explain many of its **properties**.
Examiners like asking you to relate something's structure to its properties, so make sure you're clear on this.

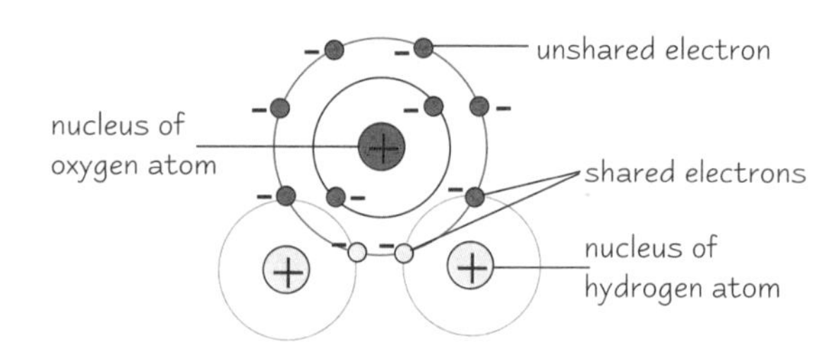

Water is **one atom of oxygen** joined to **two atoms of hydrogen** by **shared electrons**. Because the shared hydrogen electrons are pulled close to the oxygen atom, the other side of each hydrogen atom is left with a **slight positive charge**. The unshared electrons on the oxygen atom give it a **slight negative charge**. That means water is a **polar** molecule — it has negative charge on one side and positive charge on the other.

The **negatively charged oxygen atoms** of water **attract** the **positively charged hydrogen atoms** of other water molecules. This attraction is called **hydrogen bonding** and it gives water some of its special properties.

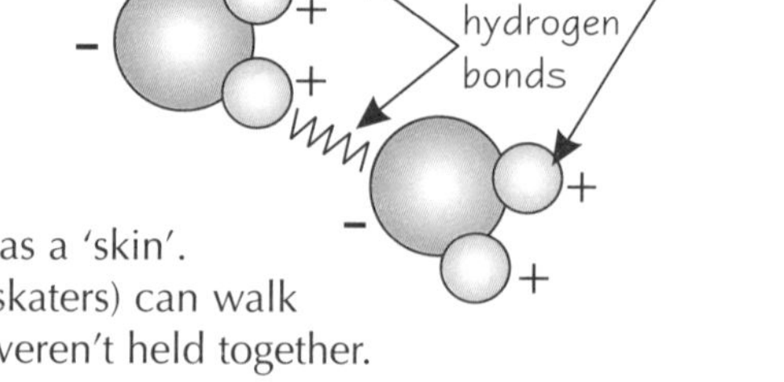

1) **Cohesion** — water molecules tend to stick together. This property enables water to flow.
2) **Surface tension** — water behaves as though it has a 'skin'. That's why some small invertebrates (like pond-skaters) can walk on water. It wouldn't happen if the molecules weren't held together.
3) **High specific heat capacity** — specific heat capacity is the **energy required** (in joules) to raise the temperature of 1 gram of a compound by 1°C. Water has a high specific heat capacity — it takes a lot of energy to heat it up. This is useful for living organisms (which all contain a high % of water) because it stops rapid temperature changes. This means they can keep their temperature fairly stable.
4) **High latent heat of evaporation** — it takes a lot of heat to evaporate water, so it's great for cooling things.

Water

Water's **Polarity** Makes it a **Good Solvent**

Because water is **polar**, it's a **good solvent** for other polar molecules. **Ionic** substances like salt, and **organic** molecules that have an **ionised group** will dissolve in water. The **positive** end of a water molecule will be attracted to a **negative ion** and the negative end of a water molecule will be attracted to a **positive ion**. The ion gets **totally surrounded** by water molecules — in other words, it **dissolves**.

Remember — a molecule is polar if it has a negatively charged bit and a positively charged bit. This is also called "uneven charge distribution."

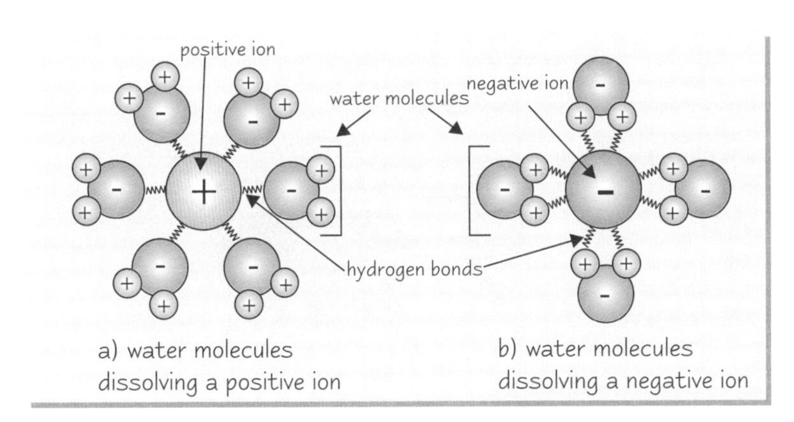

Chemical reactions take place much more easily in solution, because the dissolved ions are more **free to move around** and react than if they were held tightly together in a solid. Once dissolved, the solute can **easily be transported** by the water.

Ice Floats — Which is **Useful**

Another weird property of water that's useful in nature is that it reaches its **maximum density** at about **4°C**. This means that ice is **less dense** than the water around it, so it **floats**. This acts as an **insulation** for the water below — so the sea or a lake won't freeze solid, which allows organisms under the ice to survive.

Practice Questions

Q1 State four uses of water in living organisms.

Q2 What is a "polar molecule"?

Q3 Define "specific heat capacity".

Q4 Does water have a high or low latent heat of evaporation?

Q5 State two reasons why water is useful as a solvent in living systems.

Q6 Explain why the fact that ice floats on water is useful to living organisms.

Exam Question

Q1 Relate the structure of the water molecule to its uses in living organisms. [12 marks]

Psssssssssssssssssssssssssssssssssssss — need the loo yet?

Water is pretty darn useful really. It looks so, well, dull— but in fact it's scientifically amazing, and essential for all kinds of jobs — like maintaining aquatic temperatures, transporting things and enabling reactions. You need to learn all its properties and uses. Right, I'm off — when you gotta go, you gotta go.

Cells, Tissues and Organs

Woohoo — cells. Well, we're all made of cells, so you can't knock 'em really.

There are Two Types of Cell — Prokaryotic and Eukaryotic

PROKARYOTES	EUKARYOTES
Extremely small cells (0.5-3.0 μm diameter)	Larger cells (20-40 μm diameter)
No nucleus — DNA free in cytoplasm	Nucleus present
Cell wall made of a polysaccharide, but not cellulose or chitin	Cellulose cell wall (in plants and algae) or chitin cell wall (in fungi)
Few organelles	Many organelles
Small ribosomes	Larger ribosomes
Example: *E. coli* bacterium	Example: Human liver cell

Plasmids are small round pieces of DNA in the cytoplasm

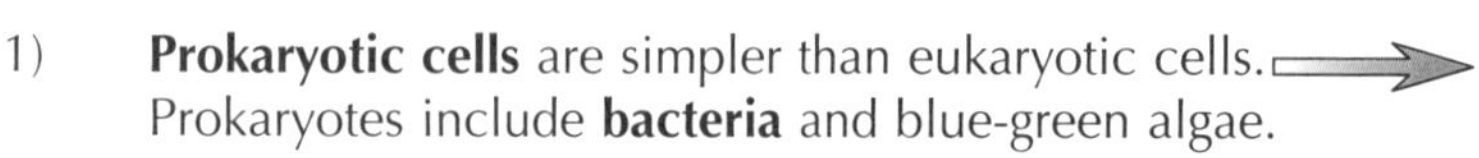

1) **Prokaryotic cells** are simpler than eukaryotic cells. Prokaryotes include **bacteria** and blue-green algae.
2) **Eukaryotic cells** are more complex, and include **all animal** and **plant cells**. The next few pages cover eukaryotic cells.

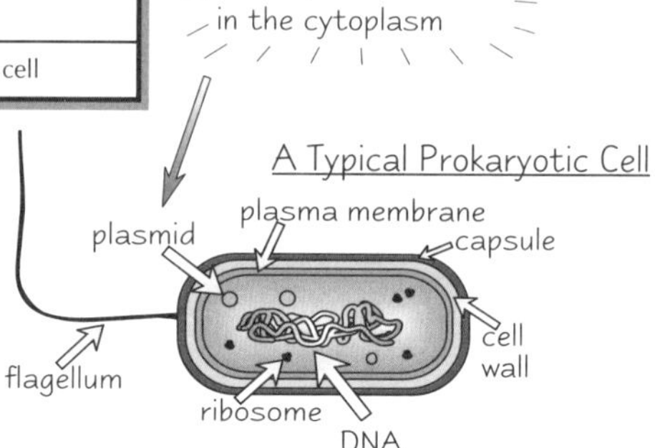

Some Cells are Adapted for Specific Functions

Most **eukaryotic cells** in multicellular organisms are **adapted** to do a particular job. 'Fraid you need to know some examples.

1) **Epithelium cells** in the **small intestine** are adapted to **absorb food efficiently**. The walls of the small intestine have lots of finger-like projections called **villi** to increase surface area — and the cells on the surface of the villi have their own **microvilli** to increase surface area even more. They also have lots of **mitochondria** — to provide energy for the active transport of digested food molecules into the cell.

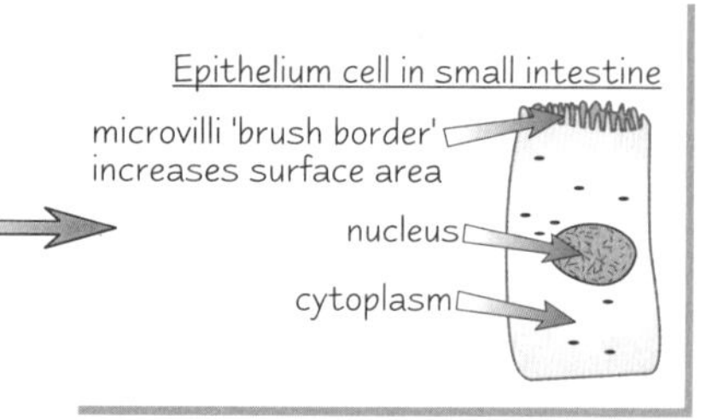

2) **Palisade mesophyll cells** in leaves do most of the photosynthesis. They contain **many chloroplasts**, so they can absorb as much sunlight as possible. The cells are **long and thin** — so lots of cells can be fitted in close together, to absorb lots of light. They also have a **thin cell wall** so that there is a short diffusion pathway for carbon dioxide into the cell (carbon dioxide is needed for photosynthesis).

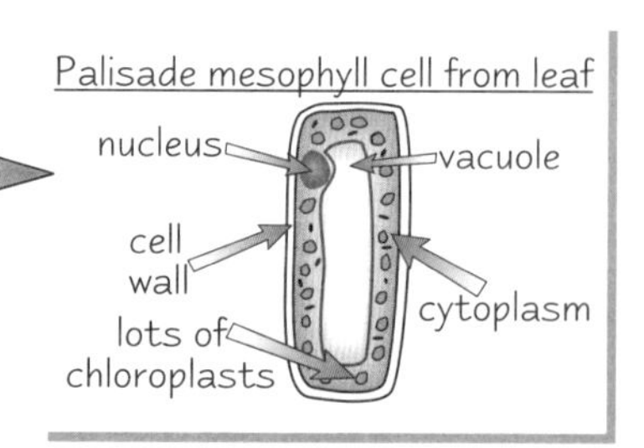

Cells are Organised into Tissues and Organs

A single-celled organism performs all its life functions in its one cell. **Multicellular organisms** (like us) are more complicated — different cells do different jobs, so cells have to be **organised** into different groups.

1) Similar cells are grouped together into **tissues**, e.g. **muscle**, **blood** and **epithelial** tissue.
2) A **group of tissues** that works together to perform a **particular function** is called an **organ**, e.g. the liver, kidneys and pancreas are organs.

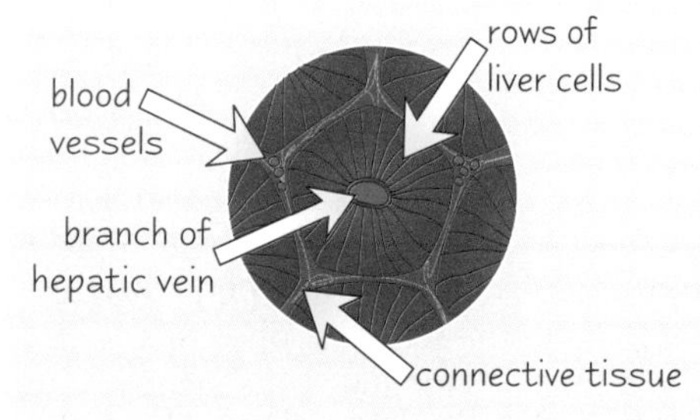

The **liver** is an example of an animal organ.
1) **Liver cells** are the main tissue.
2) There are blood vessels containing blood to provide food and oxygen for the liver cells. **Blood** is a tissue (yes, really).
3) **Connective tissue** holds the organ together.

Blood vessels aren't a tissue, though. They contain several tissues (epithelium, muscle etc.), so they're actually organs.

Organelles

Cells Contain Structures Called *Organelles*

ORGANELLE	DIAGRAM	STRUCTURE	FUNCTION
Cell wall	plasma membrane, cell wall, cytoplasm	A rigid structure that surrounds **plant cells**. It's made mainly of the carbohydrate **cellulose**.	**Supports** plant cells.
Plasma membrane	plasma membrane, cytoplasm	The membrane found on the surface of **animal cells** and just inside the cell wall of **plant cells**. It's made mainly of **protein** and **lipids**.	**Regulates the movement** of substances into and out of the cell. It also has **receptor molecules** on it, which allow it to respond to chemicals like hormones.
Nucleus	nuclear membrane, nucleolus, nuclear pore, chromatin	A large organelle surrounded by a **nuclear membrane**, which contains many **pores**. The nucleus contains **chromatin** and often a structure called the **nucleolus**.	The **chromatin** contains the genetic material (DNA) which **controls the cell's activities**. The pores allow substances (e.g. RNA) to move between the nucleus and the cytoplasm. The **nucleolus** makes **RNA**.
Ribosome	small subunit, large subunit	A **very small organelle** either floating free in the cytoplasm or attached to rough endoplasmic reticulum. Made of one large subunit and one small one.	The **site** where **proteins** are made.
Rough Endoplasmic Reticulum (RER)	ribosome, fluid	A system of membranes enclosing a fluid-filled space. The surface is **covered with ribosomes**.	**Transports proteins** which have been made in the ribosomes.
Smooth Endoplasmic Reticulum		Similar to rough endoplasmic reticulum, but with **no ribosomes**.	**Transports lipids** around the cell.
Golgi Apparatus (or 'Golgi Body')	vesicle	A group of smooth endoplasmic reticulum consisting of a series of **flattened sacs**. Vesicles are often seen at the edges of the sacs.	**Packages** substances that are produced by the cell, mainly proteins and glycoproteins. It also **makes lysosomes**.
Mitochondrion	outer membrane, inner membrane, crista, matrix	They are usually oval. They have a **double membrane** — the inner one is folded to form structures called **cristae**. Inside is the **matrix**, which contains enzymes involved in respiration (but, sadly, no Keanu Reeves).	The **site of respiration**, where **ATP** is produced. They are found in large numbers in cells that are very active and require a lot of energy.
Chloroplast	stroma, double membrane, granum (plural = grana), lamella (plural = lamellae)	A small, **flattened** structure found in **plant cells**. It's surrounded by a **double membrane**, and also has membranes inside called **thylakoid membranes**. These membranes are stacked up in some parts of the chloroplast to form **grana**. Grana are linked together by lamellae — thin, flat pieces of thylakoid membrane.	The **site** where **photosynthesis** takes place. The light-dependent reaction of photosynthesis happens in the **grana**, and the light-independent reaction of photosynthesis happens in the **stroma**.

Microscopes

You can't get away from microscopes in biology. So you need to learn this page, otherwise all those colourful splodges will remain meaningless for ever more.

Electron Microscopes *Show More Detail than* ***Light Microscopes***

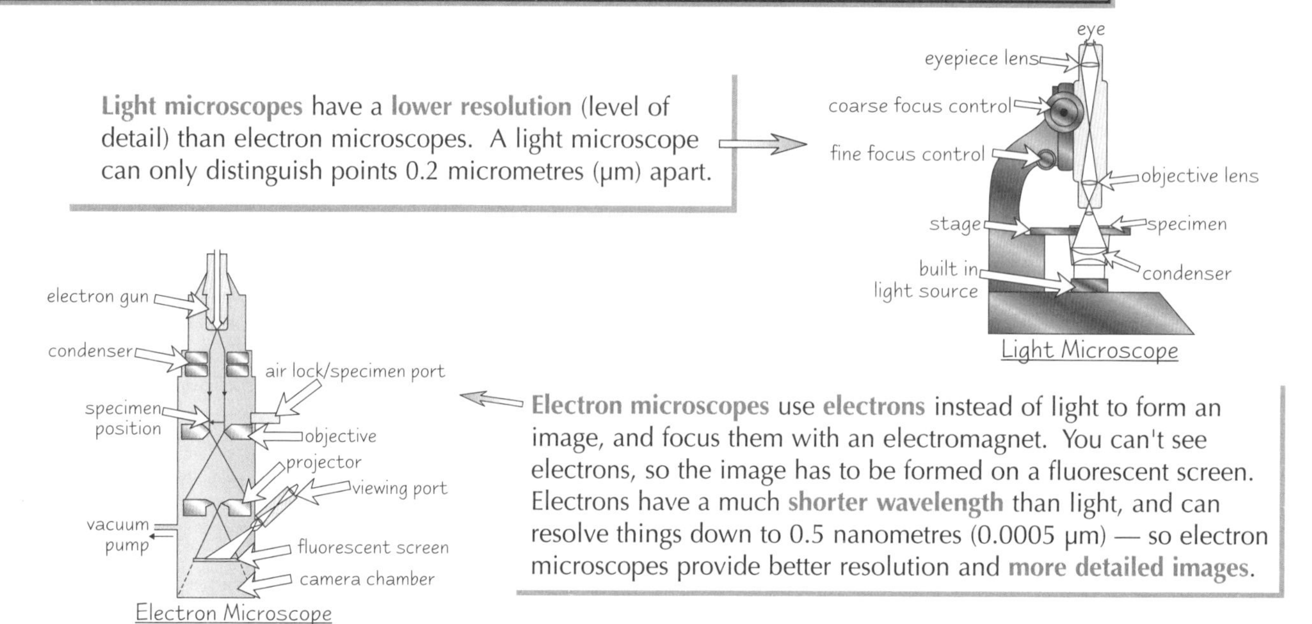

Light microscopes have a **lower resolution** (level of detail) than electron microscopes. A light microscope can only distinguish points 0.2 micrometres (μm) apart.

Electron microscopes use **electrons** instead of light to form an image, and focus them with an electromagnet. You can't see electrons, so the image has to be formed on a fluorescent screen. Electrons have a much **shorter wavelength** than light, and can resolve things down to 0.5 nanometres (0.0005 μm) — so electron microscopes provide better resolution and **more detailed images**.

Light *Microscopes Show* ***Cell Structure***

If you just want to see the **general structure of a cell**, then light microscopes are fine. But even the best light microscopes can't see most of the organelles in the cell. You can see the larger organelles, like the nucleus, but none of the internal details.

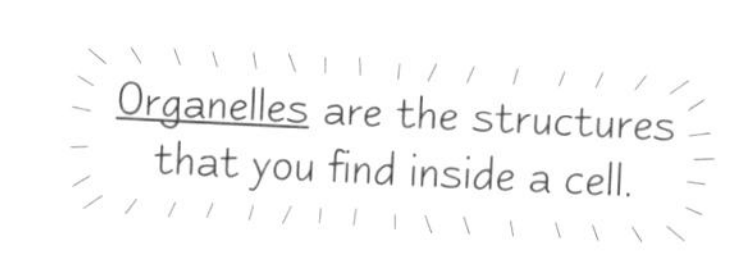

Liver cells seen under a light microscope:

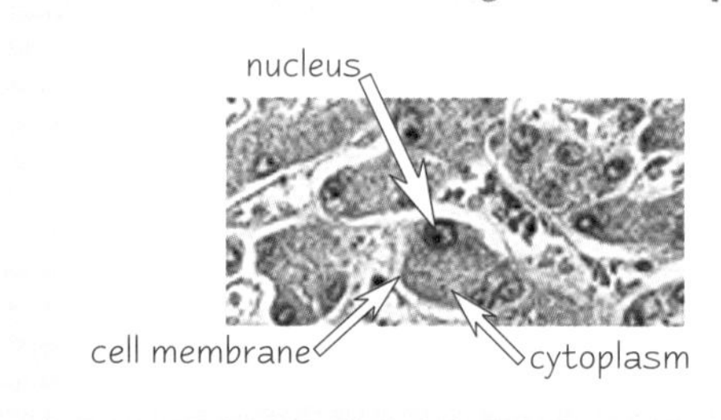

General animal cell as if seen with a light microscope

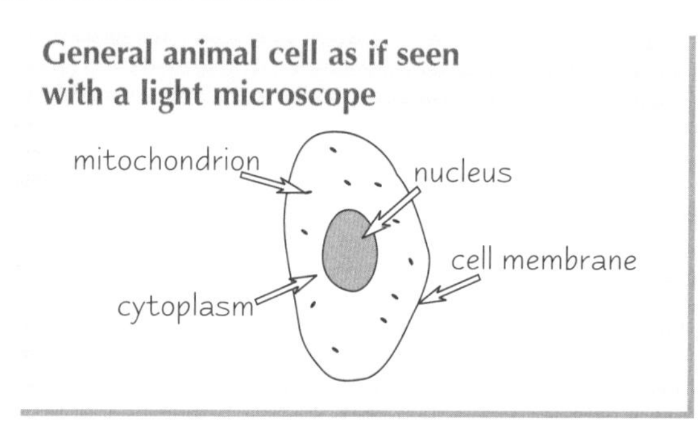

Electron *Microscopes Show* ***Organelles***

There's not much in a cell that an electron microscope can't see. You can see the **organelles** and the **internal structure** of most of them. Most of what's known about cell structure has been discovered by electron microscope studies. The diagram to the right shows what you can see in an animal cell under an electron microscope. Very pretty indeed.

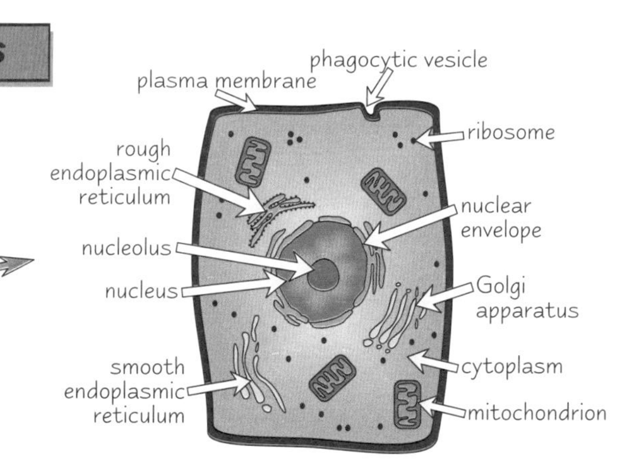

Differential Centrifugation

Differential Centrifugation Sorts Organelles

To separate a particular organelle from all the others, you use a technique called **differential centrifugation** (also called **ultracentrifugation**).

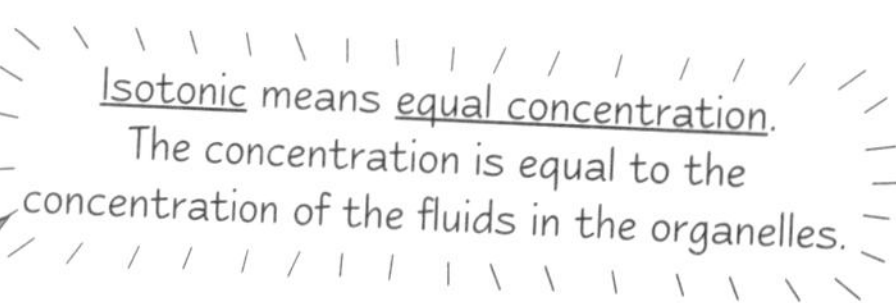

1) First, the cells are '**homogenised**' (mashed up) in **ice-cold isotonic buffer solution**. The **low temperature** prevents protein-digesting enzymes **digesting** the organelles. The **buffer** keeps the **pH constant** and the **isotonic solution** stops the organelles taking in lots of water via **osmosis** and bursting.
2) The cell fragments are poured into a **tube**. The tube is put into a **centrifuge** (with the **bottom** of the tube facing **outwards**, so heavy stuff gets flung outwards and ends up at the bottom of the tube), and is spun at a **low speed**. **Cell debris**, like the cells walls in plant cells, gets flung to the bottom of the tube by the centrifuge. It forms a thick sediment at the bottom, which is called a **pellet**. The rest of the organelles stay suspended in the **supernatant** (the fluid above the sediment).
3) The supernatant is **drained off**, poured into **another tube**, and spun in the centrifuge at a **higher speed**. The heavier organelles like the nuclei form a pellet at the bottom of the tube. The supernatant containing the rest of the organelles is drained off and spun in the centrifuge at an even higher speed.
4) This process is repeated at higher and higher speeds, until all the organelles are **separated out**. Each time, the pellets at the bottom of the tube are made up of lighter and lighter organelles.
5) The organelles which have been separated out can then be **examined** using an **electron microscope**.

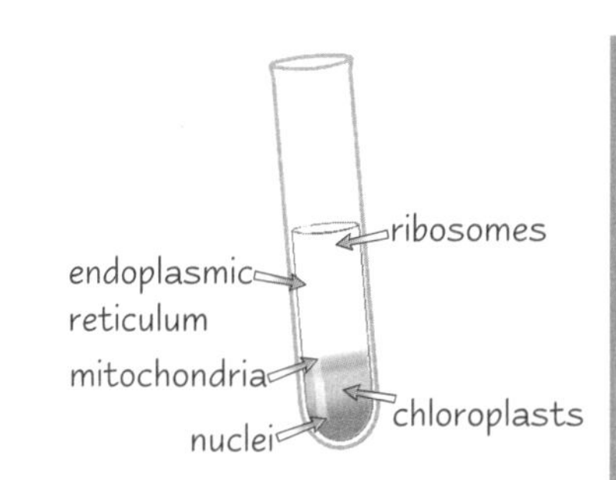

The organelles are separated in order of mass (from heaviest to lightest) — this order is usually: nuclei, then chloroplasts, then mitochondria, then endoplasmic reticulum, then finally ribosomes.

Practice Questions (Pages 12-15)

Q1 Name two organelles found only in plant cells.

Q2 Name two organelles found only in animal cells.

Q3 Name two organelles that can be seen by light microscopes, and two that can't be seen.

Q4 State the name of the process that's used to separate organelles.

Exam Questions (Pages 12-15)

Q1 The presence and number of specific organelles can give an indication of a cell's function.
Give THREE examples of this, naming the organelles concerned and stating their function. [9 marks]

Q2 a) Identify these two organelles seen in an electron micrograph, from the descriptions given below.

(i) A sausage-shaped organelle surrounded by a double membrane.
The inner membrane is folded and projects into the inner space, which is filled with a grainy material.

(ii) A collection of flattened membrane 'bags' arranged roughly parallel to one another.
Small circular structures are seen at the edges of these 'bags'. [2 marks]

b) State the functions of the two organelles that you have identified. [2 marks]

Q3 Describe how cells are prepared for differential centrifugation. [5 marks]

Learn to use a microscope — everything will become clearer...

You need to know about all the different organelles on page 13 — and be able to identify them in an electron micrograph. Differential centrifugation might have possibly one of the most poncy names in biology (and that's saying something) — but you've still got to learn all about it, otherwise you can say bye bye to easy marks.

Plasma Membranes

Two pages all about cell membranes and what they're made of. Try and contain your excitement when you read about the fluid mosaic structure — there have been some nasty cases of extreme over-excitement in the past.

Membranes Control What **Passes Through** Them

Cells and many of the **organelles** inside them are surrounded by **membranes**. Membranes have a **range of functions**:

1) Membranes control **which substances enter and leave** a cell or organelle.
2) Membranes **recognise** specific chemical substances and other cells.
3) **Membranes around organelles** divide the cell up into **different compartments** to make the different **functions more efficient** — e.g. the substances needed for **respiration** (like enzymes) are kept together inside **mitochondria**.

Cell Membranes have a **'Fluid Mosaic' Structure**

The **structure** of all **membranes** is basically the same. They are composed of **lipids** (mainly phospholipids), **proteins** and **carbohydrates** (usually attached to proteins or lipids).

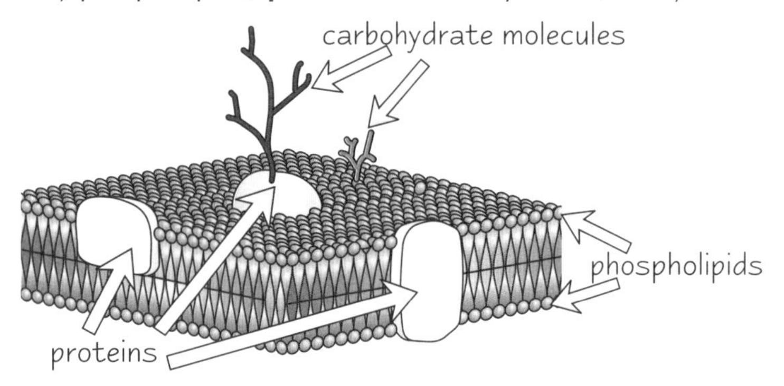

In 1972, the **fluid mosaic model** was suggested to describe the arrangement of molecules in the membrane. In the model, **phospholipid molecules** form a continuous, double layer (**bilayer**). This layer is 'fluid' because the phospholipids are constantly moving. **Protein molecules** are scattered through the layer, like tiles in a **mosaic**.

Detailed 3D pictures of cell membranes support the fluid mosaic model. Also, experiments with cell fusion show that proteins move about in the membrane, which means that the membrane is fluid.

Phospholipids Can Form **Bilayers**

Phospholipids consist of a **glycerol molecule** plus **two molecules** of **fatty acid** and a **phosphate group** (see p.7).

1) The phosphate / glycerol head is **hydrophilic** — it attracts water. The **fatty acid tails** are **hydrophobic** — they repel water.
2) In **aqueous** (**water-based**) **solutions** phospholipids automatically arrange themselves into a **double layer** so that the **hydrophobic tails** pack together **inside the layer** away from the water, and the **hydrophilic heads face outwards** into the aqueous solutions.

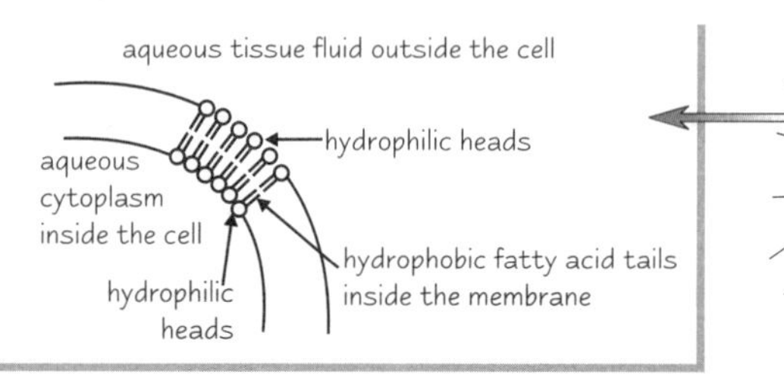

Water soluble molecules (e.g. glucose) can't pass through the fatty, hydrophobic interior of the membrane.

Plasma Membranes

Intrinsic *and* ***Extrinsic*** *Proteins Have* ***Different Functions*** *in the* ***Membrane***

1) **Intrinsic** proteins **completely span** the membrane from inside to outside.
2) **Extrinsic** proteins only **partly** span the membrane — they're stuck in either the **outer** phospholipid layer or the **inner** phospholipid layer.
3) Intrinsic and extrinsic proteins have **different functions** — usually, intrinsic proteins are for **transport** and extrinsic proteins are **receptors**.
4) **Intrinsic channel proteins** (pores) form a tiny **gap** in the membrane to allow water soluble molecules and ions through by diffusion.
5) **Intrinsic carrier proteins** carry water soluble molecules and ions through the membrane by **active transport** and **facilitated diffusion** (see p.20).

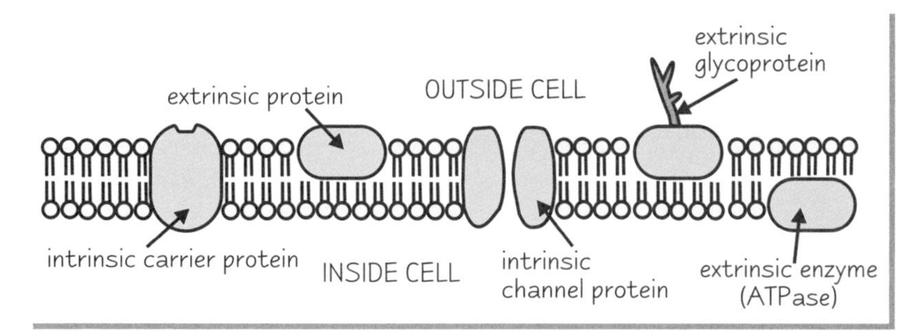

6) **Extrinsic** proteins recognise and bind on to **specific molecules** (e.g. hormones).
7) **Enzymes** can be embedded in the inner membrane of a cell or organelle — e.g. ATPase in the inner membrane of mitochondria.
8) Proteins in the membrane also help **strengthen** the membrane. There are **hydrogen bonds** between the proteins and the hydrophilic heads of the phospholipids.

Practice Questions

Q1 Give three functions of cell membranes.

Q2 How does the fluid mosaic model suggest that membranes are "fluid" and like a "mosaic"?

Q3 Which part of a phospholipid molecule is hydrophobic?

Q4 Which part of a phospholipid molecule is hydrophilic?

Q5 Name two molecules, other than phospholipids and proteins, that are present in animal cell membranes.

Q6 What types of molecule do intrinsic carrier proteins transport through the membrane?

Q7 What is the function of extrinsic proteins in the membrane?

Exam Question

Q1 a) How does a phospholipid differ from a triglyceride? [1 mark]

b) Describe the role of phospholipids in controlling the passage of water soluble molecules through the cell membrane. [2 marks]

Membranes actually ***are*** *all around...*

The cell membrane is a complex structure — but then it has to be, 'cos it's the line of defence between a cell's contents and all the big bad molecules outside. Don't confuse the cell membrane with the cell wall (found in plant cells). The cell membrane controls what substances enter and leave the cell whereas the cell wall provides structural support.

Diffusion and Osmosis

There are four methods of transport across a cell membrane. You need to learn all four — diffusion, osmosis, facilitated diffusion and active transport. It's a big topic, which is why there are four whole pages dedicated to it.

Diffusion is the Passive Movement of Particles

1) If there's a **high concentration** of particles (molecules or ions) in one area of a liquid or gas, then these particles will gradually move and **spread out** into areas of **lower concentration**. Eventually, the particles will be **evenly distributed** throughout the liquid or gas. This movement is called diffusion.
2) Diffusion is described as a **passive process** because **no energy** is needed for it to happen.
3) Diffusion can happen **across cell membranes**, as long as the particles can **move freely** through the membrane. For example, water, oxygen and carbon dioxide molecules are small enough to pass easily through pores in the membrane.

The Speed of Diffusion Depends on Several Factors

1) The **concentration gradient** is the path between an area of higher concentration and an area of lower concentration. Particles diffuse **faster** when there is a **high concentration gradient** (a big difference in concentration between the two areas).
2) The **shorter** the **distance** the particles have to travel, the **faster** the rate of diffusion.
3) **Small molecules** move faster than large molecules, so they **diffuse faster**.
4) At **high temperatures** particles have more **kinetic** (movement) energy, so they **diffuse more quickly**.
5) The larger the **surface area** of the cell membrane, the faster the rate of diffusion.

The rate at which a substance diffuses can be worked out using **Fick's law**:

$$\text{rate of diffusion} \propto \frac{\text{surface area} \times \text{difference in concentration}}{\text{thickness of membrane}}$$

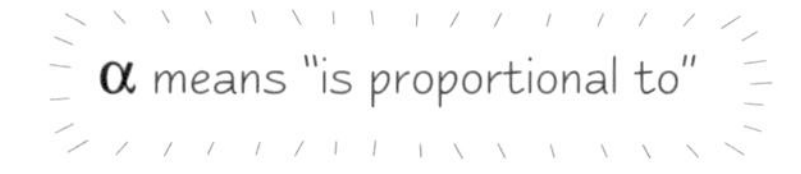

Osmosis is a Particular Kind of Diffusion

1) Osmosis is when **water molecules** diffuse through a **partially permeable membrane** from an area of **higher water potential** (i.e. higher concentration of water molecules) to an area of **lower water potential**.
2) A **partially permeable membrane** allows some molecules through it, but not all. Water molecules are small and can diffuse through easily but large solute molecules can't.
3) Water molecules will diffuse **both ways** through the membrane — but the **net movement** will be to the side with a **lower concentration of water molecules**.

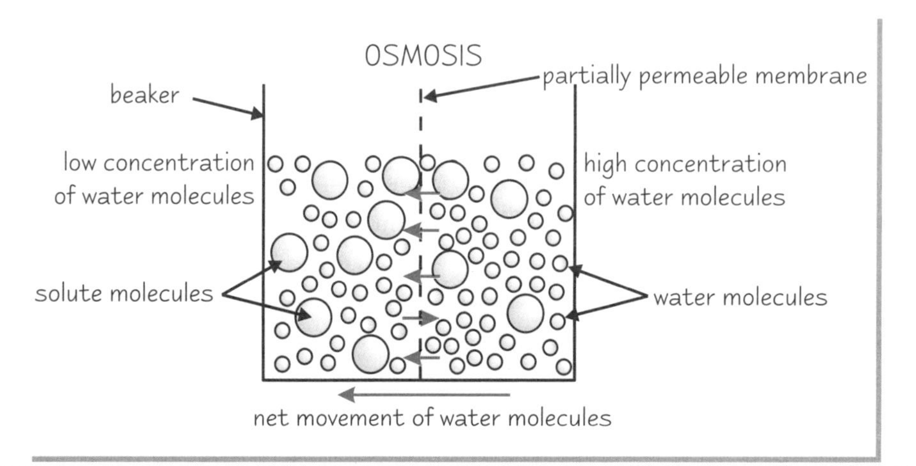

Partially permeable membranes can be useful at sea.

Diffusion and Osmosis

Water Potential is the *Ability* of Water Molecules to *Move*

1) **Water potential** is the potential (likelihood) of water molecules to diffuse out of a solution.

   Water molecules are **more likely** to diffuse out of solutions with a **higher concentration** of water molecules. These solutions have a **high water potential**.

   Water molecules are **less likely** to diffuse out of solutions with a **lower concentration** of water molecules — these have a **low water potential**.

2) Water potential is represented by the symbol ψ. It's measured in **kilopascals** (kPa).
3) **Pure water** has the **highest water potential** and is given the value of **zero kilopascals**. All solutions have a **lower** water potential than pure water, so their water potentials are always **negative**.
4) Water molecules **diffuse** from areas with a **higher water potential** to areas with a **lower water potential**.

You can *Calculate* the *Water Potential* Inside Cells

The water potential in a cell depends on **two factors**.
You can use these factors to calculate water potential. Clever.

1) **SOLUTE POTENTIAL** (ψ_s)

   The amount of **solute molecules** in a solution affects its water potential. Solute molecules form **weak chemical bonds** with water molecules and slow down their movement.
   The amount by which solute molecules **reduce** the water potential of a solution is called its **solute potential**.

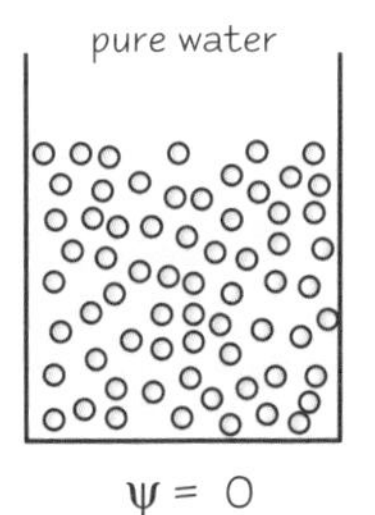

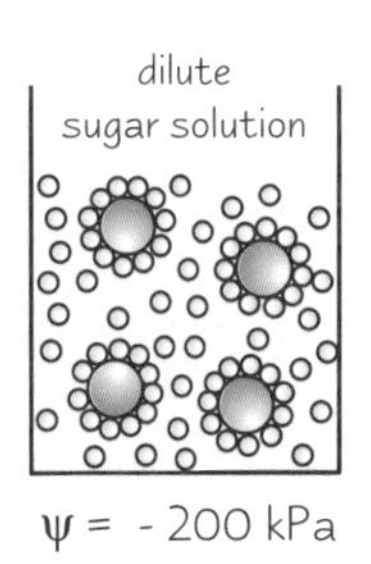

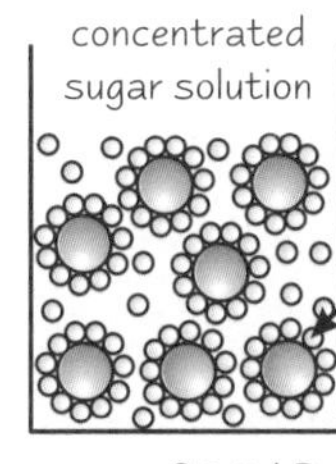

2) **PRESSURE POTENTIAL** (ψ_p)

   In cells the water potential is affected by the **cell membrane** (and the cell wall in plants). These **exert pressure inwards** on the cell, in effect squeezing water molecules out the cell. This pressure is called **pressure potential**.

$$\text{water potential } (\psi) = \text{solute potential } (\psi s) + \text{pressure potential } (\psi p)$$

Solutions can be *Isotonic*, *Hypotonic* or *Hypertonic*

Isotonic, **hypotonic** and **hypertonic** are terms that describe how the **solute potentials** of solutions **compare** with each other.

Solutions which have the **same solute potential** are **isotonic**. Their water potentials are therefore the same, so if they're separated by a partially permeable membrane, there is **no net movement** of water between the two.

A **hypotonic** solution has a **lower solute potential**, and therefore a **higher water potential**, than another solution. So there would be a net movement of water **from** the hypotonic solution to the other solution through a partially permeable membrane.

A **hypertonic solution** has a **higher solute potential** and **lower water potential** than another solution.
The net movement of water across a partially permeable membrane would be **into** the hypertonic solution.

Ginantonic solution — my gran's favourite...

*A good way to describe **osmosis** is that it's the diffusion of water molecules through a partially permeable membrane from an area of **higher water potential** to an area of **lower water potential**. Water potential is a tricky idea — but it impresses the examiners, so try and get your head round it.*

Osmosis and Facilitated Diffusion

And there's more...

Cells are Affected by the Water Potential of the Surrounding Solution

You need to learn the diagrams below for what happens to plant and animal cells when they're put into solutions of different concentrations. Flick back a page to see how it's linked to water potential.

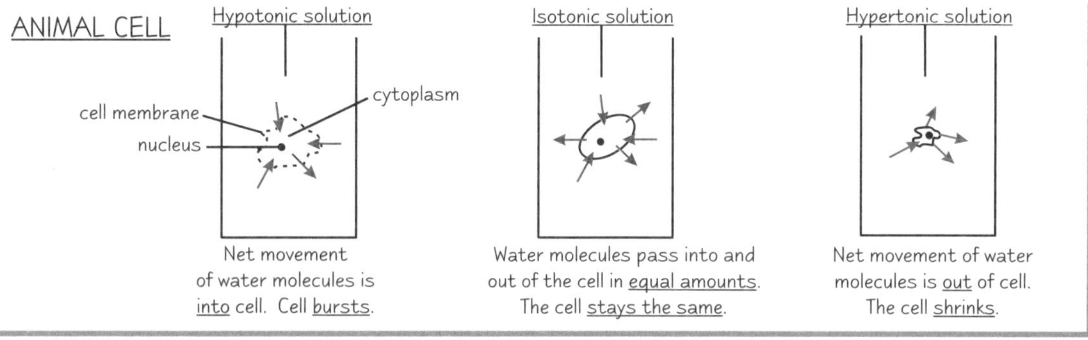

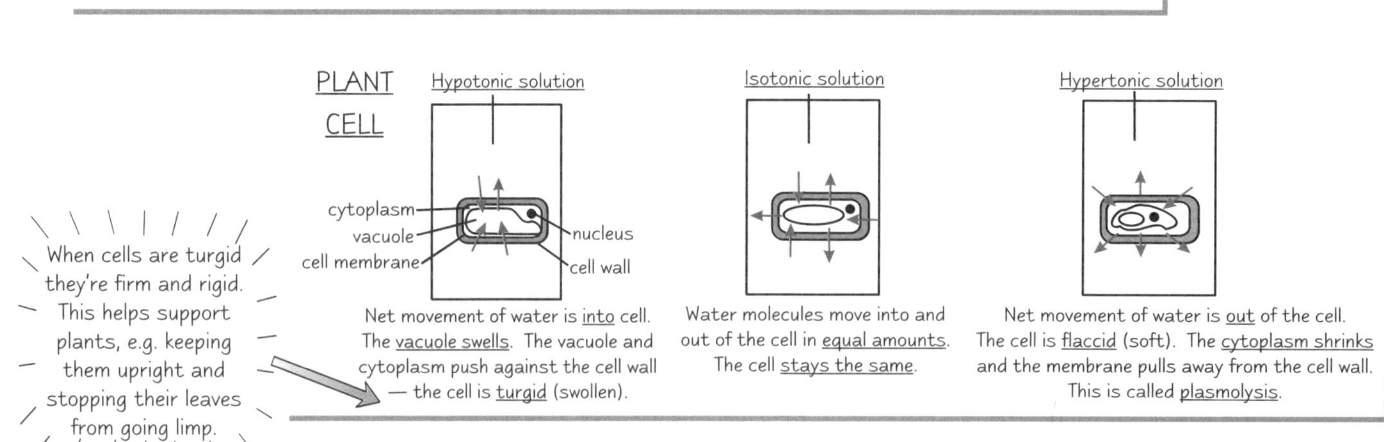

Facilitated Diffusion uses Carrier Proteins and Channel Proteins

Some **larger molecules** (e.g. amino acids, glucose) and **charged atoms** (e.g. sodium ions) can't diffuse through the phospholipid bilayer of the cell membrane themselves. Instead they diffuse through **carrier proteins** or **channel proteins** in the cell membrane. This is called **facilitated diffusion**.

1) Channel proteins form **pores** through the membrane for charged particles to diffuse through.
2) Carrier proteins **change shape** to move large molecules into and out of the cell:

The carrier proteins in the cell membrane have **specific shapes** — so specific carrier proteins can only facilitate the diffusion of specific molecules. Facilitated diffusion can only move particles along a **concentration gradient**, from a higher to a lower concentration. It **doesn't** use any **energy**.

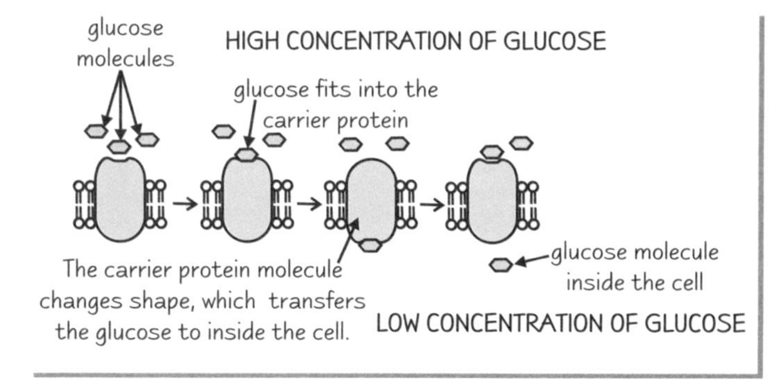

Active Transport

Active Transport Moves Substances **Against** a Concentration Gradient

1) Active transport uses **energy** to move **molecules** and **ions** across cell membranes, **against** a **concentration gradient**.
2) Molecules attach to **specific carrier proteins** (sometimes called 'pumps') in the **cell membrane**, then **molecules of ATP** (adenosine triphosphate) provide the energy to change the shape of the protein and move the molecules across the membrane.
3) Cells which carry out active transport have lots of **mitochondria** to **provide the ATP** needed.
4) Things which **reduce** the **rate of respiration**, also reduce the **rate of active transport** — because there is less energy available, e.g. lower oxygen levels or lower temperature.

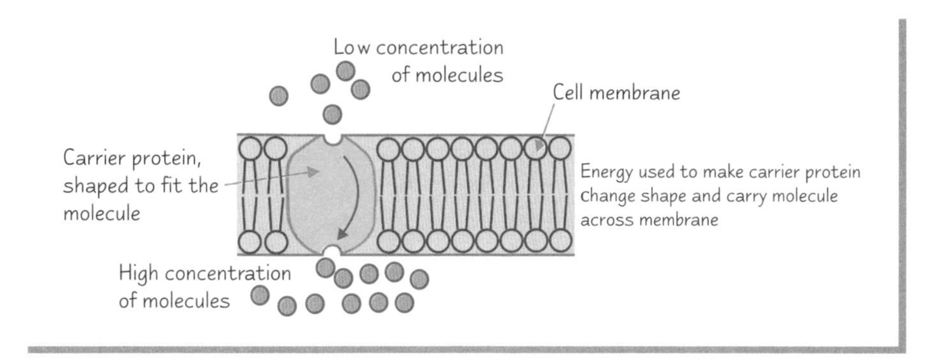

Practice Questions (Pages 18-21)

Q1 What is meant by the term "concentration gradient"?

Q2 What is a partially permeable membrane?

Q3 Define water potential.

Q4 Active transport and facilitated diffusion both involve carrier proteins. Which one needs energy?

Q5 Write down two factors which affect the rate of active transport.

Exam Questions (Pages 18-21)

Q1 The effect of temperature on the rate of diffusion of sodium ions was investigated. Pieces of potato, of equal surface area, were placed in distilled water at different temperatures. After 20 minutes the concentration of sodium ions in the water was measured and the results used to plot this graph.

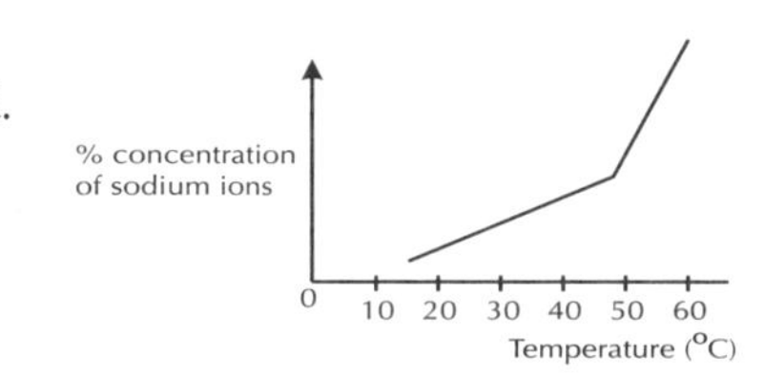

Explain the increase in the rate of diffusion between 20°C and 40°C. [2 marks]

Q2 In terms of water potential, explain how water moves from the soil into a root hair cell. [3 marks]

I tried to revise biology but I needed a higher concentration gradient...

Phew, the end of a mammoth topic on transport through the cell membrane — so now you can move on and forget it ever happened. Just kidding (I should be doing stand-up, no really) — now you need to go back over it and check you know the details. Learn the differences between similar terms, like hypertonic and hypotonic, and diffusion and facilitated diffusion.

Exchange Surfaces

Gases, nutrients and waste get into and out of organisms across exchange surfaces, which are usually cell membranes. Big organisms need to exchange a lot of stuff — and have developed specialised exchange surfaces for the purpose.

Smaller Animals have Higher Surface Area : Volume Ratios

A mouse has a bigger surface area **relative to its volume** than a hippo. This can be hard to imagine, but you can prove it mathematically. Imagine these animals as cubes:

The hippo could be represented by a block with an area of 2 cm × 4 cm × 4 cm.

Its **volume** is 2 × 4 × 4 = **32 cm³**

Its **surface area** is 2 × 4 × 4 = 32 cm² (top and bottom surfaces of cube)
+ 4 × 2 × 4 = 32 cm² (four sides of the cube)

Total surface area = **64 cm²**

So the hippo has a **surface area : volume ratio** of 64 : 32 or **2 : 1**.

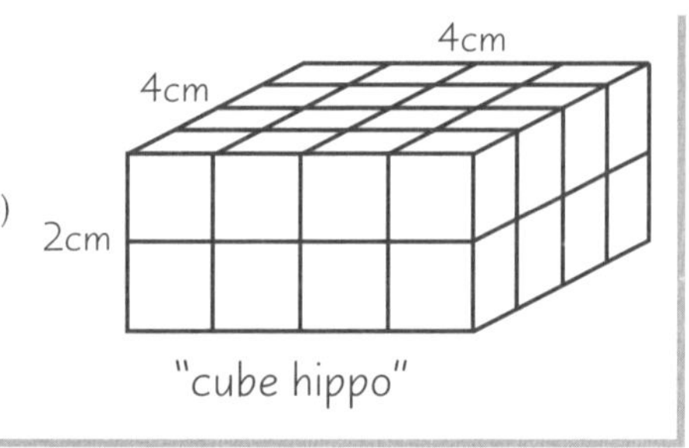

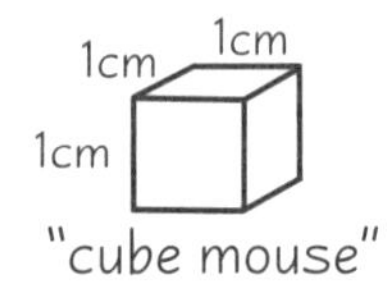

Compare this to a mouse cube measuring 1 cm × 1 cm × 1 cm

Its **volume** is 1 x 1 x 1 = **1 cm³**

Its **surface area** is 1 x 1 x 6 = **6 cm²**

So the mouse has a **surface area : volume ratio** of 6 : 1

The cube mouse's surface area is six times its volume. The cube hippo's surface area is only twice its volume. Smaller animals have a bigger surface area compared to their volume.

Organisms Need to Exchange Materials with the Environment

Cells need oxygen (for aerobic respiration) and nutrients. They need to excrete waste products like CO_2.

1) Microscopic **one-celled organisms** have very **high** surface area : volume ratios, so they can exchange materials with the environment over their whole surface by diffusion.
2) Larger multicellular organisms have **lower** surface area : volume ratios. Therefore they need specialised exchange surfaces with **big surface areas** and **specialised cells** for exchange (e.g. lungs). Many have also developed **transport systems** like blood, to carry gases, nutrients and wastes to and from inner cells.

Surface area is also important for **body temperature**. Animals release heat energy when their **cells respire** and lose it to the **environment**. So **small animals** with high S.A. : volume ratios, like mice, have to use lots of energy just **keeping warm**, while **big animals** with low S.A. : volume ratios, like elephants, are more likely to **overheat**.

Elephants have a low S.A. : volume ratio, but their big, flat ears have a very small volume and a big surface area. Elephants use their ears for heat exchange, to help heat escape from the body quickly.

Organisms need to get Gases To and From their Gas Exchange Surfaces

Organisms which have **specialised gas exchange surfaces** (e.g. lungs), need a **ventilation system** — a way to get **oxygen** from the surrounding **environment** to these surfaces, and to get **carbon dioxide** away from these surfaces back out into the environment. In mammals, the ventilation system is **breathing**:

Gas	% Inhaled	% Exhaled
Oxygen	20.96	16.50
Carbon Dioxide	0.04	4.00
Nitrogen	79.00	79.00
Water Vapour	Variable	Saturated

The table above shows the difference in composition between the air you breathe in and the air you breathe out again a moment later — the main difference is you've used up some oxygen and added a bit of CO_2.

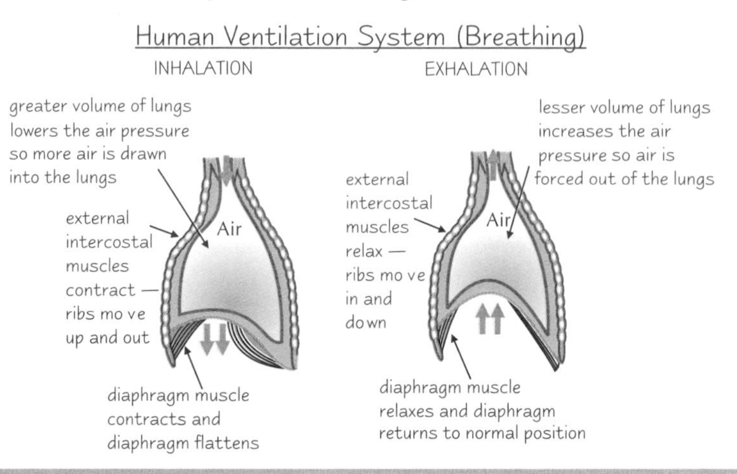

Exchange Surfaces

Lungs are Specialised Internal Organs for Gas Exchange

Mammals, like humans, exchange oxygen and carbon dioxide through their **lungs**. The lungs have special **features** that make them well-adapted for **breathing**:

Cartilage — rings of strong but bendy cartilage keep the **trachea** open.

Goblet cells — produce **mucus** to trap inhaled dust and other particles.

Cilia — **hairs** on the cells that line the trachea, bronchi and bronchioles. They **move** to push the mucus with trapped particles **upwards**, away from the lungs.

Smooth muscle — round the bronchi and bronchioles. **Involuntary** muscle contractions narrow the airways.

Elastic fibres — between the alveoli. Stretch the lungs when we breathe in and recoil when we breathe out to help push air out.

Pleural membrane — protective lining on the lungs.

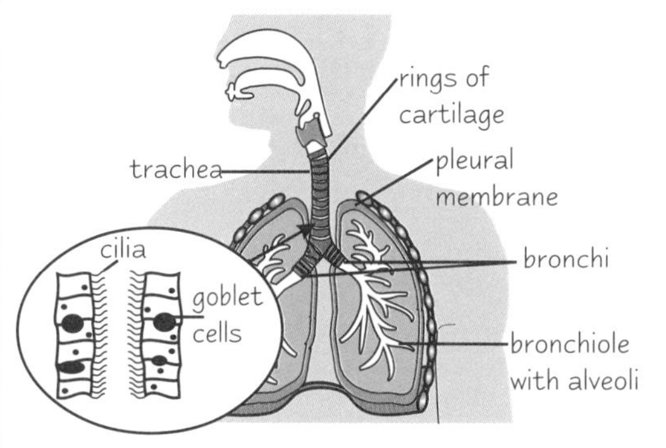

In Humans, Gas Exchange Happens in the Alveoli

Lungs contain millions of microscopic air sacs called **alveoli**, which are responsible for gas exchange. They're so tiny, but so important — size ain't everything.

1) The huge number of alveoli means a **big surface area** for exchanging oxygen and carbon dioxide.
2) O_2 diffuses **out of** alveoli, across the **alveolar epithelium** (the single layer of cells lining the alveoli) and the **capillary endothelium** (the single layer of cells of the capillary wall), and into **haemoglobin** in the **blood**.
3) CO_2 diffuses **into** the alveoli from the blood, and is breathed out through the lungs, up the trachea, and out of the mouth and nose.
4) The alveoli secrete liquid called **surfactant**. This stops the alveoli collapsing by lowering the surface tension of the water layer lining the alveoli.

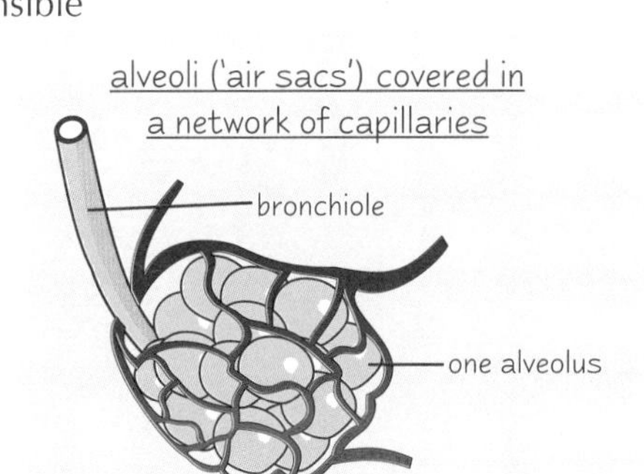

Alveoli have **adaptations** that make them a really good surface for gas exchange. They have the following features, which all **speed up** the **rate of diffusion**:

- **thin exchange surfaces** (the alveolar epithelium cells are very thin)
- **short diffusion pathways** (the alveolar epithelium layer is only one cell thick)
- **a large surface area to volume ratio**
- **a steep concentration gradient** between the alveoli and the capillaries surrounding them

Practice Questions

Q1 Why do large, multicellular organisms need specialised exchange surfaces?

Q2 How does surface area : volume ratio affect the body temperature of animals?

Q3 What is a ventilation system?

Q4 Write down four ways in which the alveoli in the lungs are adapted for gas exchange.

Exam Question

Q1 Explain why a specialised gas exchange system is required in humans. [4 marks]

Alveoli — useful things...always make me think of pasta...

I know you've just got to the end of a page — but it would be a pretty smart idea to have another look at diffusion and osmosis (pages 18-20). It's all the stuff about Fick's Law, and transport across membranes. Not the most thrilling prospect I realise, but it'll help these pages make more sense.

Gas Exchange in Fish

I know a bloke who's studying gas exchange in fish for his PhD. He also loves Dr. Who to an abnormal extent. I'm not saying the two are connected, but who knows? My advice is to learn this page, but to make darn sure you don't enjoy it. Otherwise, you'll be dressing up as K9 and buying annuals and not doing revision 'cos you're in love with Tom Baker...

Fish Use a **Counter-Current System** for Gaseous Exchange

There's a **lower concentration** of oxygen in water than in air. So **fish** have special **adaptations** to get **enough oxygen**.

1) Water containing oxygen enters the fish through its **gills**. In the gills there's a **one-way current** of water that's continually kept flowing by a **pumping mechanism**.
2) Each gill is made of many very **thin plates** called **gill filaments**, which give a **big surface area** for **exchange** of **gases**. The gill filaments are covered in lots of tiny structures called **lamellae** which **increase** the **surface area** even more. The lamellae have lots of blood capillaries and a thin surface layer of cells to speed up diffusion.
3) In the gills, the **blood** flows over the lamellae in one direction and **water** flows over the lamellae in the opposite direction. This is called a **counter-current system**. It means that a **concentration gradient** is maintained between the water and the blood — so as much oxygen diffuses from the water into the blood as possible.

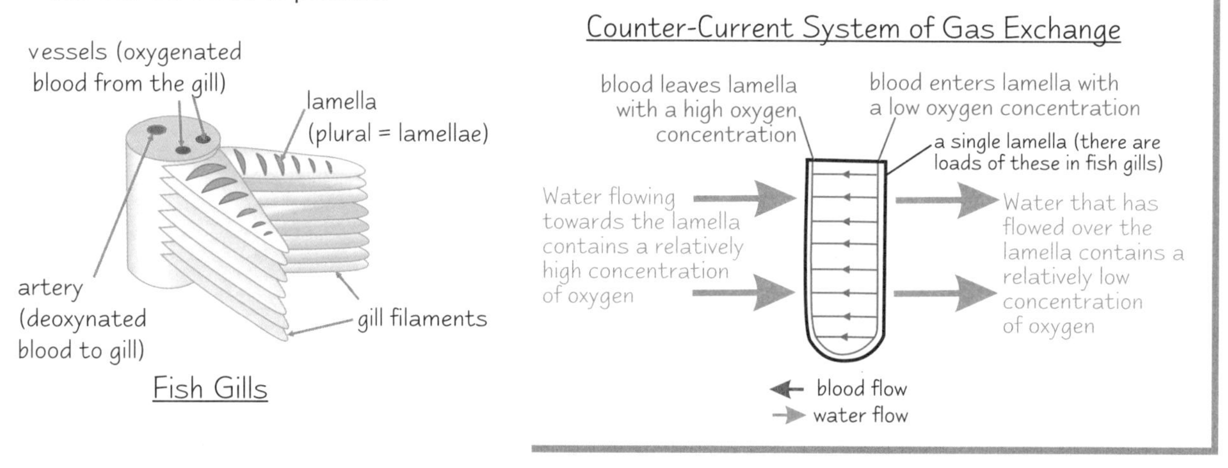

Bony Fish have a **Ventilation System** too

Fish have to be able to get **water** (containing **oxygen**) from the surrounding environment to their exchange surfaces (the gills). To do this they have a **ventilation system**.

1) The fish **opens its mouth** and lowers the floor of the **buccal cavity** (the inside of its mouth).
2) This means the **volume** of the fish's mouth increases and the **pressure** in it decreases — so **water is drawn in**.
3) The **opercular cavity** (don't ask me...see the diagram) gets **bigger** and the fish **closes its mouth** and raises the floor of the buccal cavity. So the pressure in the buccal cavity increases and the pressure in the opercular cavity decreases.
4) This means that water is forced from the buccal cavity **over the gills** and into the opercular cavity.
5) The water is finally **expelled** through the opercular valve.

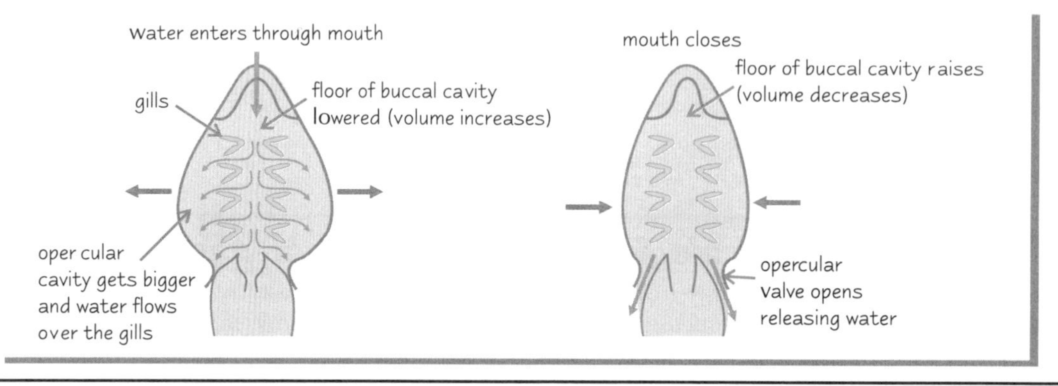

Gas Exchange in Plants

Plants Exchange Gases at the Surface of the Mesophyll Cells

Two processes in plants involve gas exchange:

1) **Respiration takes in oxygen** and **releases carbon dioxide** — just like in animals.
2) **Photosynthesis** involves **taking in carbon dioxide** and **releasing oxygen**.

The main gas exchange surface is the **surface of the spongy mesophyll cells** in the leaf. This is well adapted for its function:

1) It's **moist** which means gases diffuse faster.
2) There's a **large surface area** for gas exchange.
3) Mesophyll cells have **thin cell walls** and membranes — so there is a **short diffusion pathway** for gases into the cell.
4) There are **lots of air spaces** in the spongy mesophyll layer which allow gases to circulate.

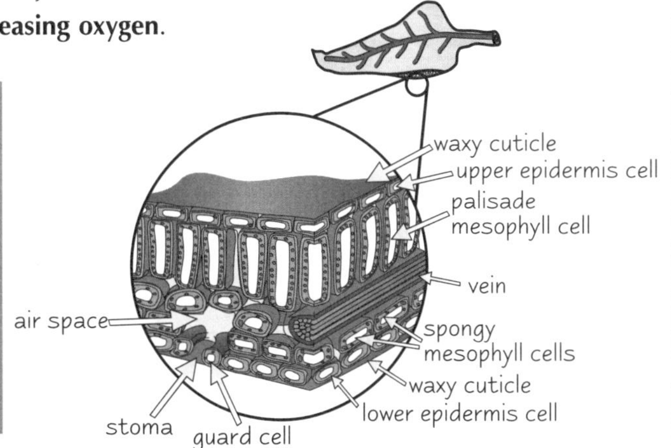

Gases Enter and Leave the Leaf through Special Pores Called Stomata

The mesophyll cells are inside the leaf. Gases diffuse in and out of the leaf through special pores in the **epidermis** called **stomata** (singular = stoma). The stomata can **open** to allow exchange of gases, and **close** if the plant is losing too much water.

Each stoma (pore) is surrounded by **guard cells**. The stoma opens when the guard cells increase in **turgidity**, by absorbing water by osmosis. When this happens, each guard cell changes shape because its **inner cell wall** is **thicker** than its outer one. This opens the pore, as the diagram shows.

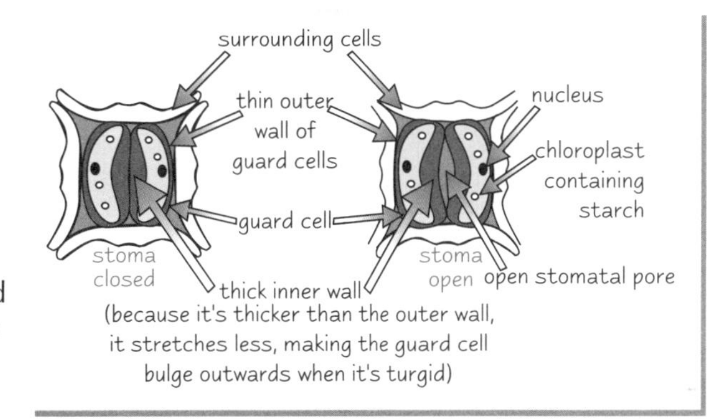

Practice Questions

Q1 What are the thin plates in the gills called?

Q2 Why is the gas exchange system at the gills of fish described as 'counter-current'?

Q3 Where is the buccal cavity in a fish?

Q4 What two processes in a plant involve gas exchange?

Q5 What is the name of the pores through which gases enter a leaf?

Exam Questions

Q1 Describe how gaseous exchange in fish is adapted to cope with the relatively low oxygen levels in water. [4 marks]

Q2 Write down four ways in which the spongy mesophyll layer in the leaf is adapted to gas exchange. [4 marks]

The Real Justin Timberlake — saucy facts REVEALED...

Justin Timberlake's been calling me all day wanting to have an informed discussion about ventilation in bony fish. I had an answerphone message this morning begging me to talk biology to him. He said my speech about the spongy mesophyll layer just wasn't enough, he needed to know about guard cells too. And instead I'm writing this book for you... dammit.

Enzymes

*Enzymes crop up loads in biology — they're really useful 'cos they make reactions work more quickly. So, whether you feel the need for some speed or not, read on — because you **really** need to know this basic stuff about enzymes.*

Enzymes are Biological Catalysts

A catalyst is a substance that speeds up a chemical reaction without being used up in the reaction itself.

Enzymes speed up chemical reactions by acting as **biological catalysts**.

1) They catalyse every **metabolic reaction** in the bodies of living organisms. Even your **phenotype** (physical appearance) is down to enzymes that catalyse the reactions that cause growth and development.
2) Enzymes are **globular proteins** (see p.5) although some have **non-protein components** too.
3) Every enzyme has an area called its **active site**. This is the part that connects the enzyme to the substance it interacts with, which is called the **substrate**.
4) Enzymes are **highly specific**. This is because of their structure as **proteins**. The **active site** of an enzyme is formed by its **tertiary structure**. If the tertiary structure is damaged (e.g. by temperature or pH changes) then the active site is also damaged and the enzyme is **denatured**.

Enzymes Reduce Activation Energy

In a chemical reaction, a certain amount of energy needs to be supplied to the chemicals before the reaction will start. This is called the **activation energy** — it's often provided as **heat**. Enzymes **reduce** the amount of activation energy that's needed, often making reactions happen at a **lower temperature** than they could without an enzyme. This **speeds** up the **rate of reaction**.

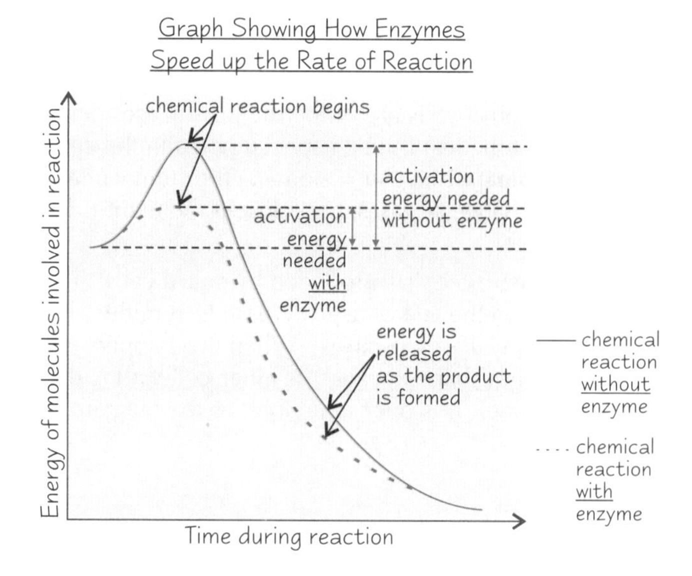

When a substrate fits into the enzyme's active site it forms an **enzyme-substrate complex**:

1) If two substrate molecules need to be **joined**, attaching to the enzyme holds them **close together**, **reducing** any **repulsion** between the molecules so they can bond more easily.
2) If the enzyme is catalysing a **breakdown reaction**, fitting into the active site puts a **strain** on bonds in the substrate, so the substrate molecule **breaks up** more easily.

The 'Lock and Key' Model is one way of Explaining How Enzymes Work...

Enzymes are a bit picky. They only work with **specific substrates** — usually only one. This is because, for the enzyme to work, the substrate has to **fit** into the **active site**. If the substrate's shape doesn't match the active site's shape, then the reaction won't be catalysed. This is called the '**lock and key**' model.

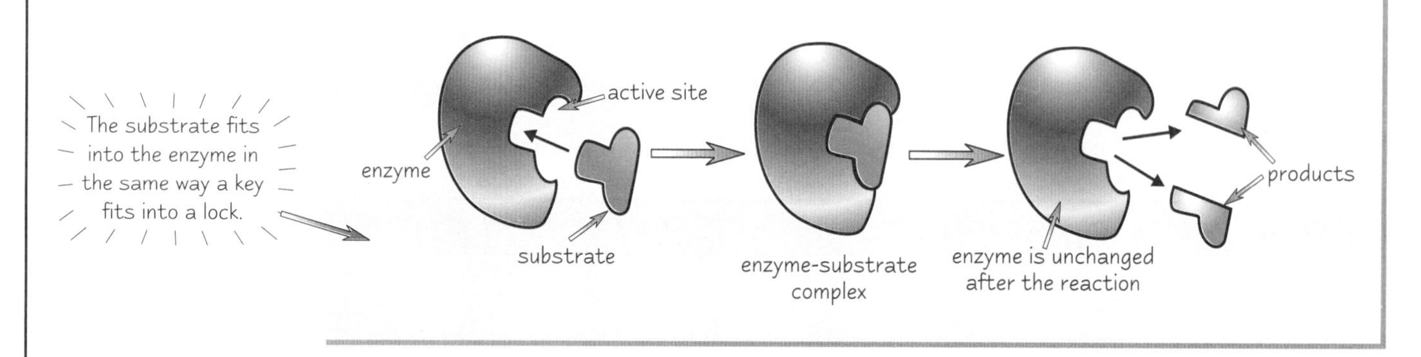

Enzymes

...but the '*Induced Fit*' Model is a *Better Theory*

Scientists now believe that the lock and key model doesn't tell the whole story.
The enzyme and substrate do have to fit together in the first place, but then it seems that the **enzyme-substrate complex changes shape** slightly to complete the fit.
This **locks** the substrate even more tightly to the enzyme. This is called the '**induced fit**' model. It helps explain why enzymes are so **specific** and only bond to one particular substrate. The substrate doesn't only have to be the right shape to fit the active site, it has to make the active site **change shape** in the right way as well.

The 'Luminous Tights' model was popular in the 1980s but has since been found to be grossly inappropriate.

Temperature has a *Big Influence* on Enzyme Activity

Like any chemical reaction, the rate of an enzyme-controlled reaction increases when the temperature's raised. More heat means more **kinetic energy**, so molecules move faster. This makes the enzyme more likely to **collide** with the substrate. But, if the temperature increases beyond a certain point, the **reaction stops**. This is because the rise in temperature also makes the enzyme's particles **vibrate**:

1) If the temperature goes above a certain level, this vibration **breaks** some of the **bonds** that hold the enzyme in shape.
2) The **active site changes shape** and the enzyme and substrate **no longer fit together**.
3) At this point, the enzyme is **denatured** — it no longer functions as a catalyst.

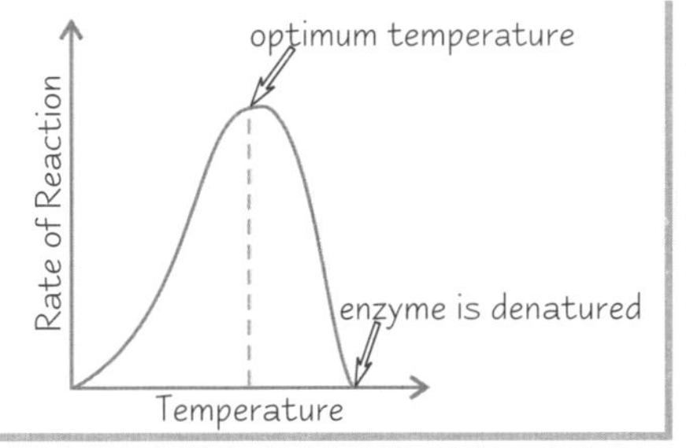

Every enzyme has an optimum temperature. In humans it's around 37°C but some enzymes, like those used in biological washing powders, can work well at 60°C.

pH Also Affects Enzyme Activity

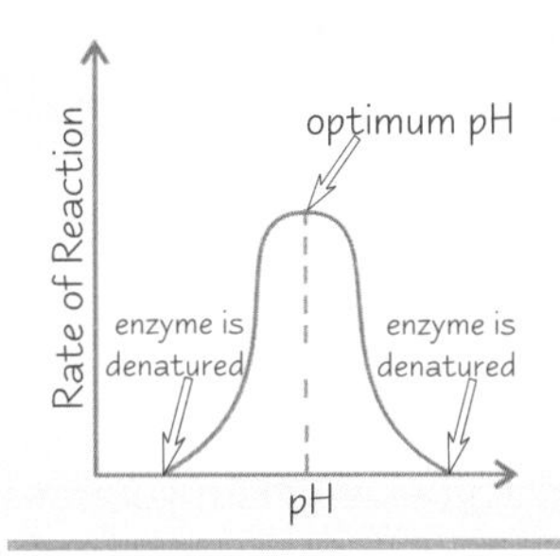

All enzymes have an **optimum pH value**. Most work best at neutral pH 7, but there are exceptions. **Pepsin**, for example, works best at acidic pH 2, which suits it to its role as a stomach enzyme. Above and below the optimum pH, the H+ and OH- ions found in acids and alkalis can mess up the **ionic bonds** that hold the enzyme's tertiary structure in place. This makes the active site change shape, so the enzyme is **denatured**.

Substrate Concentration Affects the Rate of Reaction *Up To a Point*

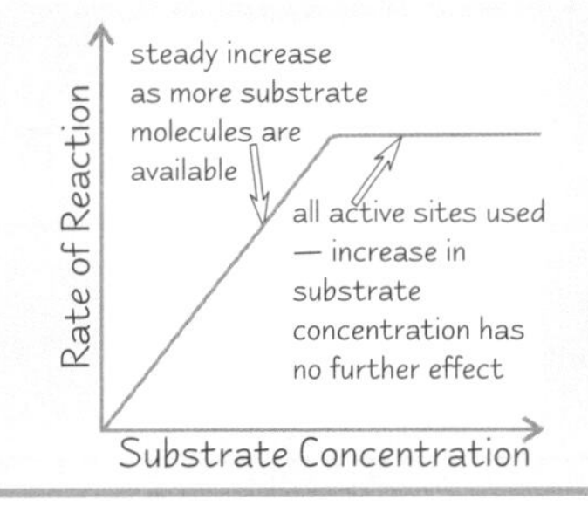

Substrate concentration affects the rate of reaction up to a certain point. The higher the substrate concentration, the faster the reaction, but only up until a **'saturation' point**. After that, there's so many substrate molecules that the enzymes have about as much as they can cope with, and adding more **makes no difference**.

This is a 3-page mini-section about enzymes — turn over for the exam questions and extraordinary witticisms.

Enzymes

Enzyme Activity can be **Inhibited**

Enzyme activity can be prevented by **enzyme inhibitors** — molecules that **bind to the enzyme** that they inhibit. Inhibition can be **competitive** (active site directed) or **non-competitive** (non-active site directed).

1) **Competitive inhibitors** have a **similar shape to the substrate**. They compete with the substrate to bond to the active site, but no reaction follows. Instead they **block** the active site, so **no substrate** can **fit** in it. How much inhibition happens depends on the **relative concentrations** of inhibitor and substrate — if there's a lot of the inhibitor, it'll take up all the active sites and stop any substrate from getting to the enzyme.

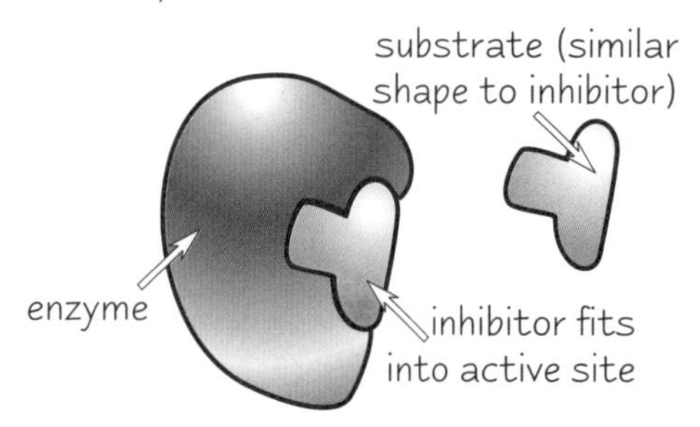

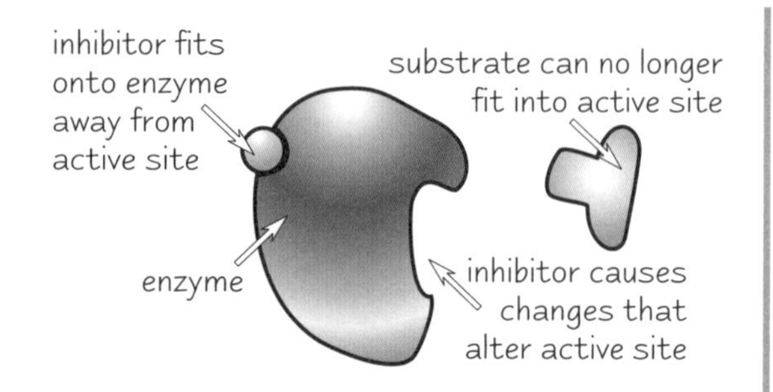

2) **Non-competitive inhibitors** bond to the enzyme **away from its active site**, but this causes the active site to **change shape**.
They don't 'compete' with the substrate because even if there's a substrate in the active site, the inhibitor can still fit on.

Practice Questions (Pages 26- 28)

Q1 Define the term "catalyst".

Q2 What is an "enzyme-substrate complex"?

Q3 Explain why enzymes are specific (i.e. only work with a single or a small group of substrates).

Q4 Will increasing the substrate concentration always increase the rate of enzyme reaction? Explain your answer.

Q5 How do high temperatures denature enzymes?

Q6 What is the difference between a competitive and a non-competitive enzyme inhibitor?

Exam Questions

Q1 When doing an experiment on enzymes, explain why it is necessary to control the temperature and pH of the solutions involved. [8 marks]

Q2 When a small amount of chemical X is added to a mixture of an enzyme and its substrate, the formation of reaction products is reduced. Increasing the amount of X in the solution causes further reduction in products. State, with reasons, the likely nature of chemical X. [4 marks]

Don't be shy — lose your inhibitions and learn these pages...

It's not easy being an enzyme. They're just trying to get on with their jobs, but the whole world seems to be against them sometimes. High temperature, wrong pH, inhibitors — they're all out to get them. Sad though it is, make sure you know every word. Learn how different factors affect enzyme activity, and be able to describe the different types of inhibitors.

Digestion and Absorption

Time for a bit of blood and guts — well, actually, mainly just the guts. A whole three pages of it — with a bit about saprophytic digestion thrown in for good measure. I'm starting to feel queasy.

The **Human Gut** is **Adapted** to its **Function**

There are two stages in human digestion:

1) **Mechanical breakdown** of large pieces of food into small pieces.
2) **Chemical breakdown** of large molecules into small molecules.

The human gut (**alimentary canal**) is composed of different parts, each with a specific job to do:

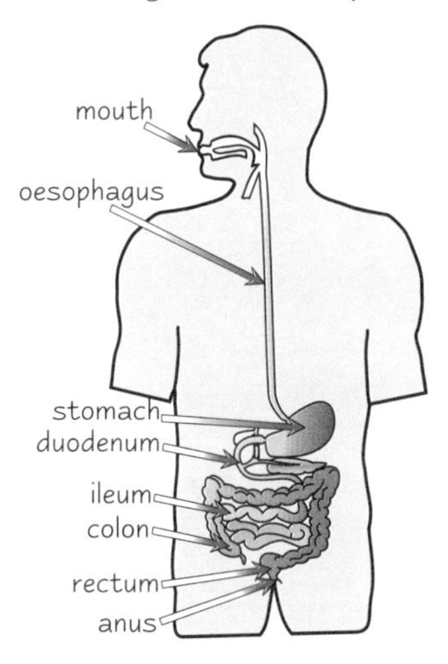

The Alimentary Canal

1) **Mouth** — **Mastication** (chewing) of food by teeth **mechanically** breaks up food so there's a **larger surface** area for enzymes to work on. Mixing food with saliva (water, **amylase** and mucus) partly digests food so it can be swallowed easily. The amylase **hydrolyses** starch into maltose.

2) **Oesophagus** — A tube that takes food from the mouth to the stomach using waves of muscle contractions called **peristalsis**. Mucus is secreted from glandular tissue in the walls, to lubricate the food's passage downwards.

3) **Stomach** — The stomach walls produce **gastric juice**, which consists of hydrochloric acid (HCl), **pepsin** (an enzyme) and mucus. Pepsin is an **endopeptidase** — it hydrolyses peptide bonds in the **middle** of polypeptide molecules (proteins), breaking them down into smaller polypeptide chains. It only works in **acidic conditions**, which are provided by the HCl. Peristalsis in the stomach turns food into an acidic fluid, called **chyme**.

4) **Duodenum** (small intestine) — Contains **alkaline bile** and **pancreatic juice**, which neutralise chyme and break food down into small, soluble molecules:

- Bile is produced in the **liver** and stored in the **gall bladder**, then enters the duodenum through the bile duct. It **emulsifies** lipids into small droplets, which speeds up hydrolysis of lipids by **pancreatic lipase**.
- Pancreatic juice contains **digestive enzymes**:
  Lipase — Hydrolyses **lipids** into fatty acids and glycerol.
  Amylase — Hydrolyses **starch** into maltose.
  Trypsin — An **endopeptidase** that hydrolyses polypeptides into smaller polypeptides.
  Exopeptidases — Hydrolyse peptide bonds found at the **end** of polypeptide chains, giving **free amino acids**.

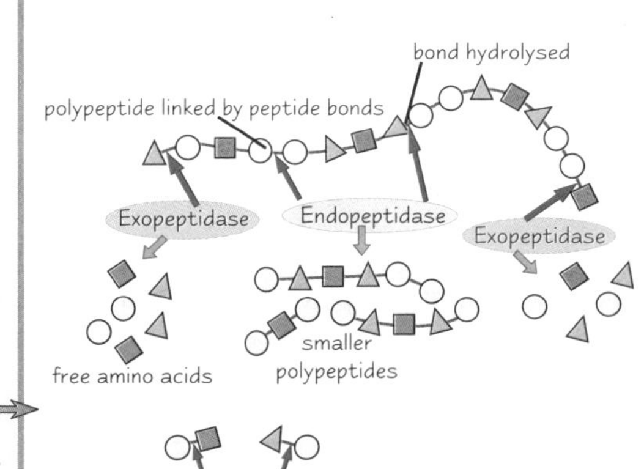

- **Intestinal juice** is produced by the gut wall. It contains more digestive enzymes — more lipases, **maltase** (hydrolyses maltose into glucose), more exopeptidases and also **dipeptidases** that hydrolyse dipeptides into amino acids.

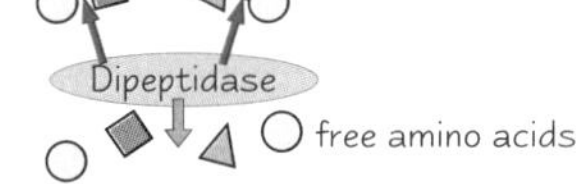

5) **Ileum** (small intestine) — The small, soluble molecules of digested food (glucose, amino acids, fatty acids and glycerol) are absorbed through the **microvilli** lining the gut wall. Absorption is through diffusion, facilitated diffusion and active transport (see p.18, 20 and 21):

Method of Absorption	Affected Molecules
Diffusion	Fatty acids, glycerol, water
Facilitated Diffusion	Some glucose, some amino acids
Active Transport	Mineral ions, some glucose, some amino acids

6) **Colon** (large intestine) — **Water** is **absorbed** by the body and waste is pushed along towards the rectum.
7) **Rectum** — Stores **faeces** until they're expelled through the **anus**.

Digestion and Absorption

The Gut Wall Consists of Four Layers of Tissue

The gut wall has the same **general structure** all the way through the human gut:

1) The **mucosa** (inner lining) lubricates the passage of food with **mucus**. This prevents **autodigestion** (enzymes attacking the gut wall). It's lined with **surface epithelium** cells.
2) The **submucosa** contains capillary beds and nerve fibres.
3) The **circular and longitudinal muscles** control the shape and movement of the gut.
4) The **serosa** contains tough tissue, which provides protection from friction against other organs.

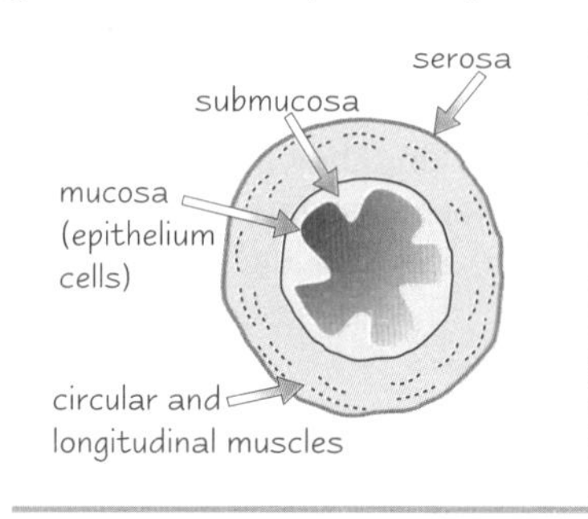

Some parts of the Alimentary Canal have Special Adaptations

The gut wall in different regions of the alimentary canal has **special features** so it can carry out specific functions:

The mucosa in the **ileum** has **villi** and **microvilli** to **increase the surface area** for absorbing the products of digestion.
It also has other adaptations for effective absorption:

- It consists of a **single layer** of epithelial cells, so there's a **short diffusion pathway**.
- A **moist lining** helps substances **dissolve** so they can pass through cell membranes.
- The **capillary bed** takes away **absorbed molecules** so the diffusion gradient is maintained.
- **Lymph vessels** take away absorbed **glycerol** and **fatty acids** to join the lymphatic system (see p.55).
  This maintains the diffusion gradient.
- **Carrier proteins** in epithelial cell membranes allow **facilitated diffusion**.
- The epithelial cells contain lots of **mitochondria** to make ATP, needed for **active transport**.

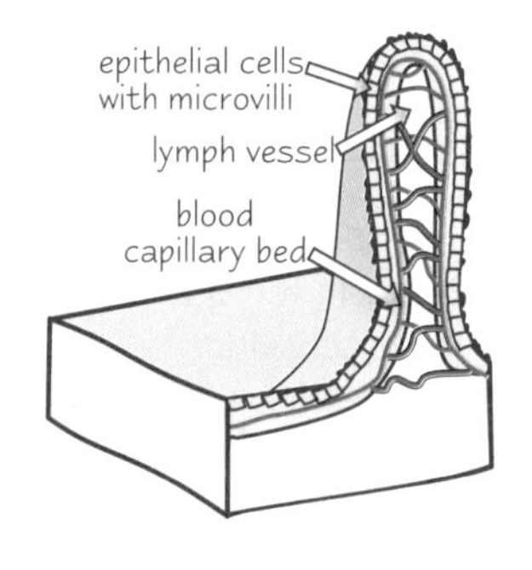

The gut wall in the **oesophagus** uses its circular and longitudinal muscles to perform **peristalsis**. These muscles work as an **antagonistic pair** — as one contracts, the other relaxes. Food is pushed along in front of the contractions.

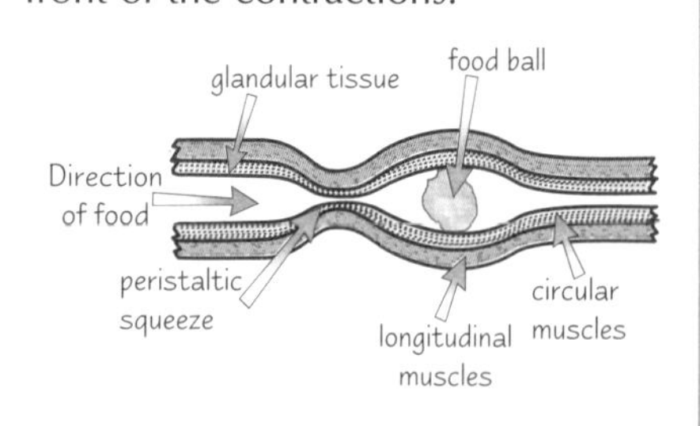

The mucosa in the **stomach** contains **gastric pits**, which secrete gastric juice.

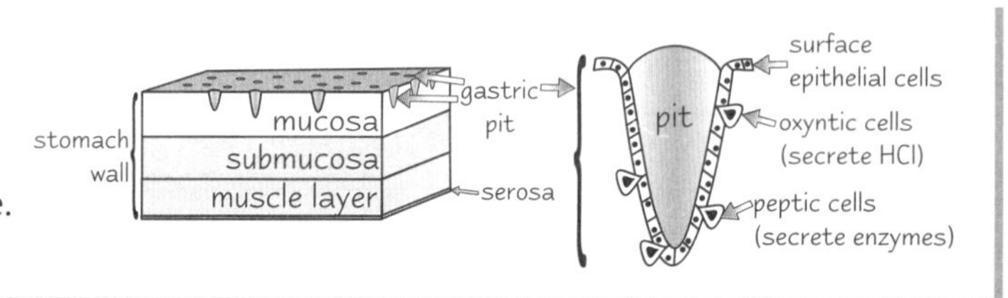

Epithelial cells in the mucosa of the **duodenum** release **digestive enzymes**.

Digestion and Absorption

Saprophytes Digest Food Outside Their Bodies

Saprophytes feed on **dead organic matter** and **wastes** like faeces.

1) **Rhizopus** is a saprophytic **fungus**, which feeds on starch and other organic materials in plants.
2) Rhizopus fungi have thread-like cells called **hyphae** to penetrate the food.
3) **Enzymes** are released from the hyphae to digest the food surrounding them. Large insoluble molecules are broken down by the enzymes, giving smaller soluble products that are then absorbed into the hyphae cells.

Saprophytic Digestion

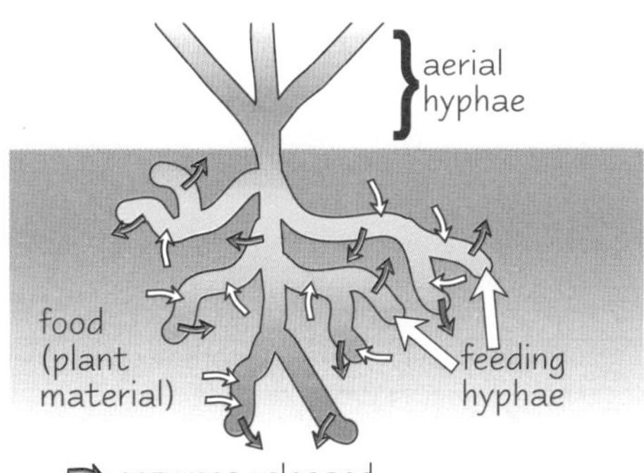

You can see evidence of extracellular saprophytic digestion by using the **starch agar assay**.

1) **Starch** and **agar** are mixed together and poured into a Petri dish.
2) **Saprophytic fungi** are grown on the agar.
3) The starch agar is flooded with **iodine** in potassium iodide solution.
4) The bits of the agar which still contain starch turn **blue-black** (see p.8). But around the area of fungal growth is a **clear zone**. This is because the starch in the agar has been **digested** by **amylase enzymes** secreted by the fungus.

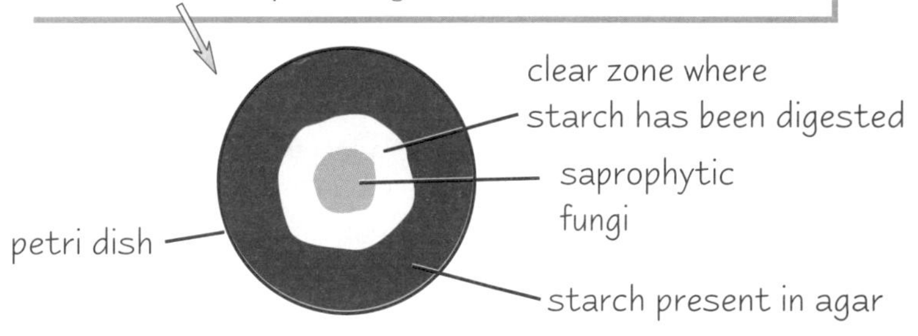

Practice Questions (Pages 29-31)

Q1 What is saliva made of?

Q2 What is an exopeptidase?

Q3 What is the difference in function between the duodenum and the ileum?

Q4 Draw the layers of the gut wall, label them and describe the functions.

Q5 Explain how peristalsis works.

Q6 Write down three ways in which the ileum is adapted to absorbing the products of digestion.

Q7 What is saprophytic digestion?

Q8 What colour does starch turn when iodine in potassium iodide solution is added to it?

Exam Questions

Q1 Explain how protein is broken down in the digestive system. [4 marks]

Q2 Describe what role the ileum plays in the human digestive system. [3 marks]

My mum told me mastication would make me go blind...

Quite a bit to learn here — it's understandable, 'cos digestion is a complicated process with many stages. The key thing is to learn what each bit of the gut does and how it's adapted to do this function. Make sure you know the difference between endo- and exopeptidases. Then eat some food, and revise its process down the body, not lingering on the last stage though.

Basic Structure of DNA and RNA

*These pages are about the structure of DNA (**deoxy**ribonucleic acid) and RNA (plain ol' ribonucleic acid), plus a little thing called DNA self replication, which is kinda important to us living things. (OK, spot the major understatement here.)*

DNA** and **RNA** are Very **Similar Molecules

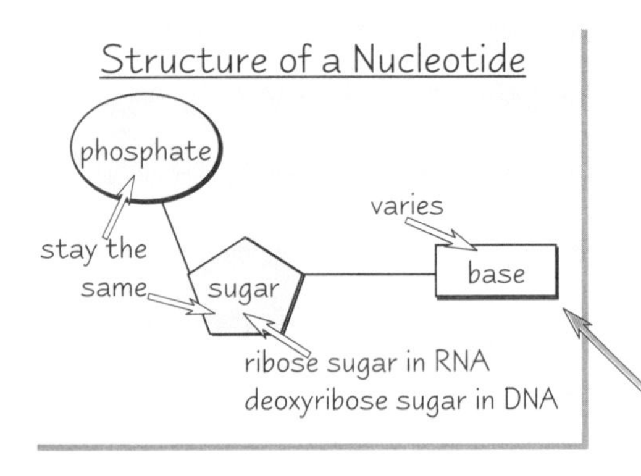

1) DNA and RNA are **nucleic acids** — they're made up of lots of **nucleotides** joined together to give **stable polynucleotides**. Nucleotides are units made from a **pentose sugar** (with 5 carbon atoms), a **phosphate** group and a **base** (containing nitrogen and carbon).

2) The sugar in **DNA** nucleotides is a **deoxyribose** sugar — in **RNA** nucleotides it's a **ribose** sugar. Within DNA and RNA, the sugar and the phosphate are the same for all the nucleotides. The only bit that's different between them is the **base**. There are five possible bases and they're split into two groups:

GROUP	BASE	found in DNA	found in RNA
Purine bases	**adenine**	✓	✓
	guanine	✓	✓
Pyrimidine bases	**cytosine**	✓	✓
	thymine	✓	×
	uracil	×	✓

Although DNA is called deoxyribonucleic acid, it still contains oxygen.

DNA** and **RNA** are Polymers of **Mononucleotides

Mononucleotides (single nucleotides) join together by a **condensation reaction** between the **phosphate** of one group and the **sugar** molecule of another. As in all condensation reactions, **water** is a by-product.

DNA is made of **two strands of nucleotides**. **RNA** has just the one strand. In DNA, the strands spiral together to form a **double helix**. The strands are held together by **hydrogen bonds** between the bases.

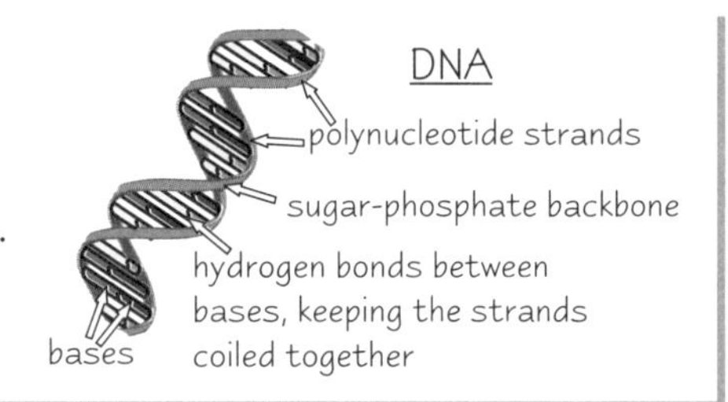

*Bonds Only Form Between **Specific Pairs** of **Bases***

Each base can only join with one particular partner — this is called **specific base pairing**.

1) In DNA, **adenine** always pairs with **thymine** (**A - T**) and **guanine** always pairs with **cytosine** (**G - C**).
2) It's the same in RNA, but **thymine**'s replaced by **uracil** (so it's **A - U** and **G - C**).

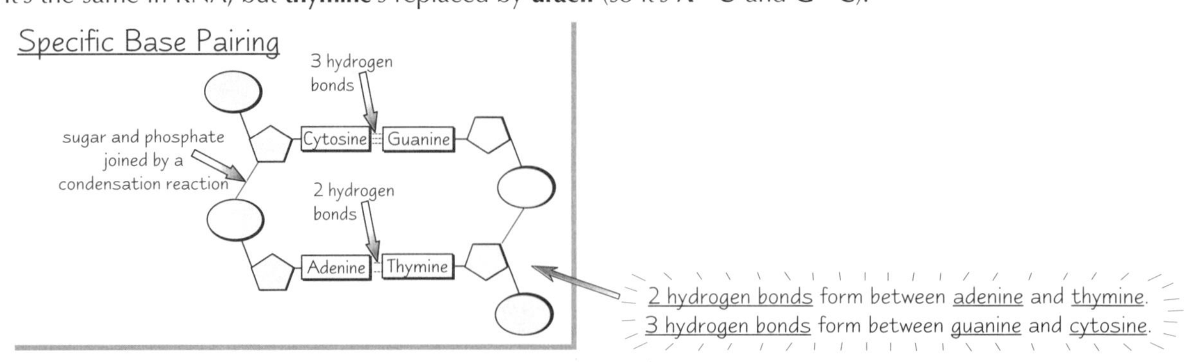

DNA's Structure** Makes it **Good at its Job

1) The job of DNA is to carry **genetic information**. A DNA molecule is very, very **long** and is **coiled** up very tightly, so a lot of genetic information can fit into a **small space** in the cell nucleus.
2) Its **paired structure** means it can **copy itself** — this is called **self-replication** (see next page). It's important for cell division and for passing on genetic information to the next generation.

Replication of DNA

DNA can Copy Itself — *Self-Replication*

DNA has to be able to **copy itself** before **cell division** can take place, which is essential for growth and development and reproduction — pretty important stuff.

1) **Specific base pairing** means that each type of base in DNA only pairs up with one other type of base — **A** with **T**, **C** with **G**. When a molecule of **DNA splits**, the **unpaired bases** on each strand can match up with complementary bases on **free-floating nucleotides** in the cytoplasm, making an **exact copy** of the DNA on the other strand. This happens with the help of enzymes. The result is **two molecules** of DNA **identical** to the **original molecule** of DNA:

Parent molecule of DNA splits.

Bases on individual free nucleotides pair up with matching bases.

The nucleotides are joined together by the enzyme DNA polymerase and hydrogen bonds form between the complementary bases on both strands.

2) This type of copying is called **semi-conservative replication** — because **half** of the new strands of DNA are from the **original** piece of DNA.

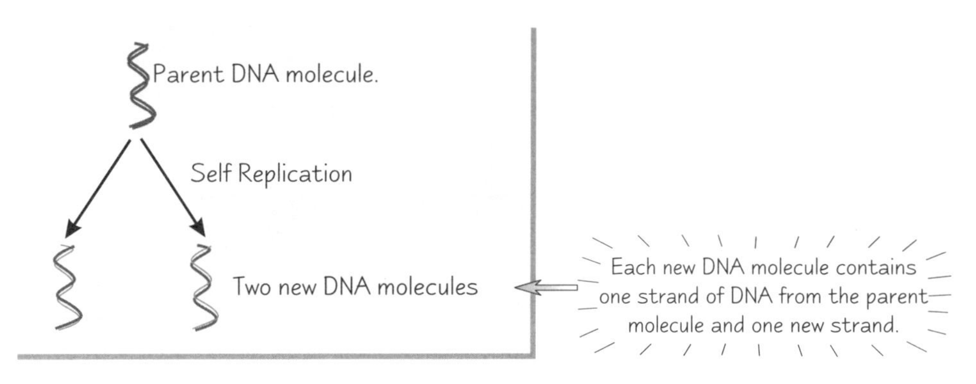

Practice Questions

Q1 What three things are nucleotides made from?

Q2 Which base pairs join together in a DNA molecule?

Q3 What type of bonds join the bases together?

Q4 What is the name used to describe the type of replication in DNA?

Exam Questions

Q1 Explain how the structure of DNA is related to its function. [2 marks]

Q2 Describe, using diagrams where appropriate, how nucleotides join together and how two single strands of DNA become joined. [5 marks]

Q3 Describe and explain the semi-conservative method of DNA replication. [6 marks]

Give me a D, give me an N, give me an A! What do you get? — very confused…

You need to know the basic structure of DNA and RNA and also how DNA's structure makes it good at its job. Then there's self replication to get to grips with — hmmm, rather you than me. I'm afraid there's nowt else you can do except buckle down, pull your socks up and get all them facts learnt.

Genes and the Genetic Code

Here comes some truly essential stuff — the genetic code is the real nitty-gritty of biology. So eyes down for some serious fact-learning. I'm afraid it's all horribly complicated — all I can do is keep apologising. Sorry.

Genes are strands of DNA

1) Genes are **lengths of DNA**. They're found on **chromosomes**.
2) **Genes** contain the genetic information that determines the development of all organisms. The **order** of the gene's **nucleotide bases** decides what the organism will be like and how it will develop.

Genes can Exist in More than One Form — called Alleles

A gene can exist in more than one form. These forms are called **alleles** — they code for **different types** of the **same characteristic**. For example, the gene that codes for **eye colour** exists as one of two alleles — one codes for the colour **blue** and the other codes for **brown.** The two alleles have **slightly different base sequences**.

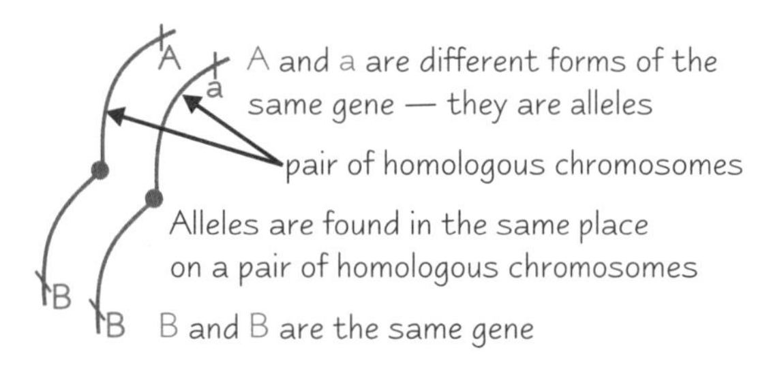

Genes are located along **chromosomes** found in the nucleus of cells. These chromosomes are paired into matching pairs (**homologous pairs**) during cell division. Alleles coding for the same characteristic will be found at the **same position** (**locus**) on each chromosome in a homologous pair. So homologous chromosomes contain the same genes, but not necessarily the same alleles.

DNA Contains the Basis of the Genetic Code

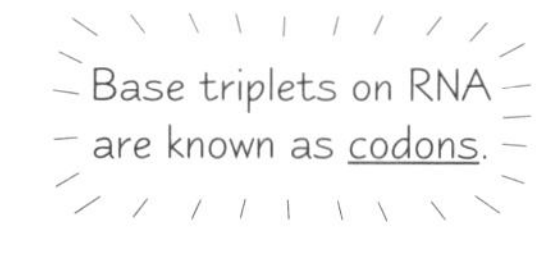

Each **gene** codes for a specific **sequence of amino acids** that forms a particular **polypeptide** (protein). The way that DNA codes for proteins is called the **genetic code**.

1) Genes code for specific amino acids with sequences of three bases, called **base triplets**. Different base triplets code for different amino acids. For example, AGA codes for serine and CAG codes for valine.
2) There are **64** possible **base triplet combinations**. There are only about **20** amino acids in human proteins so there are some base triplets to spare. These aren't wasted though:

- some amino acids use more than one base triplet.
- some base triplets act as 'punctuation' to stop and start production of an amino acid sequence. These create **stop codons** and **start codons**.

The Genetic Code is Non-Overlapping and Degenerate

1) In the genetic code, each base triplet is read in sequence, separate from the triplet before it and after it. Base triplets **don't share** their **bases** — so the code is described as **non-overlapping**.
2) The genetic code in DNA is also described as **degenerate**. This is because there are **more triplet codes** than there are amino acids. Some **amino acids** are coded for by **more than one base triplet**, e.g. Tyrosine can be coded for by TAT or TAC.

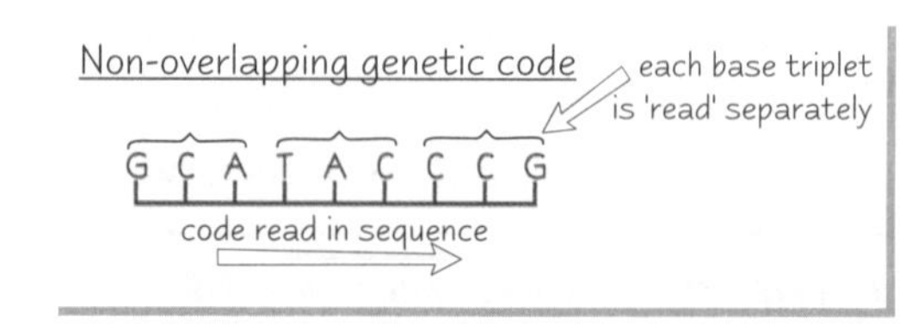

Types of RNA

There are Three Types of RNA

There are **three types** of RNA — **ribosomal RNA** (rRNA), which forms part of **ribosomes** (see page 13), and **messenger RNA** (mRNA) and **transfer RNA** (tRNA), which are both involved in **making proteins** (see over the page for more on this — bet you can't wait). You need to know details of the **structures** of **mRNA** and **tRNA**.

Messenger RNA (mRNA)

1) **mRNA** is a **single polynucleotide strand** that's formed in the **nucleus**.
2) The important thing to know about it is that it's formed by using a **gene** on a **single DNA strand** as a **template**. **Specific base pairing** means that mRNA ends up being an exact **reverse copy** of the gene (see the piccy on the right to make sense of this).
3) You also need to know that the **3 bases in mRNA** that pair up with a base triplet on the DNA strand are called a **codon**. Codons are dead important for making proteins (see p.36), so **remember this word**. Make sure you realise that a codon has the **opposite bases** to a base triplet (except the base **T** is replaced by **U** in **RNA**).

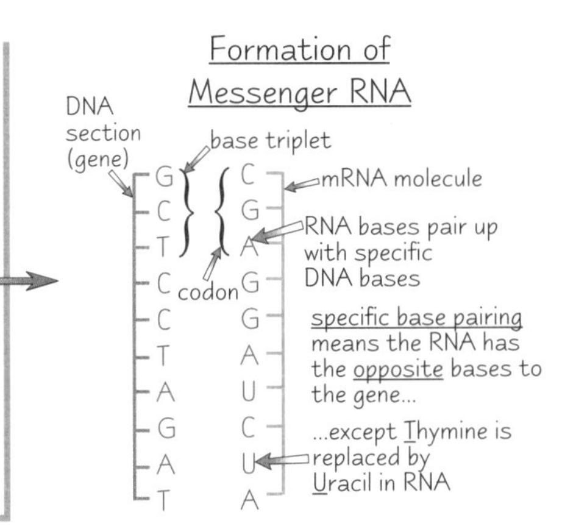

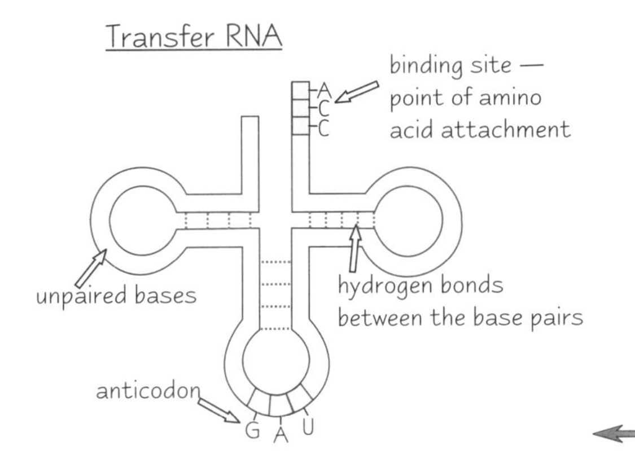

Transfer RNA (tRNA)

1) **tRNA** is a **single polynucleotide strand** that's folded into a **clover shaped molecule**.
2) Each tRNA molecule has a **binding site** at one end, where a specific **amino acid** attaches itself to the bases there.
3) Each tRNA molecule also has a specific sequence of **three bases** at one end of it, called an **anticodon**.
4) The significance of binding sites and anticodons are all revealed over the page. But you need to know **where they are found** on a tRNA molecule, so learn the diagram on the left off by heart.

Practice Questions

Q1 What is an allele?

Q2 Explain why the genetic code is described as being "degenerate".

Q3 What three types of RNA are there?

Q4 What is the name of the group of three bases on mRNA that corresponds to a base triplet on DNA?

Q5 What shape does a chain of tRNA fold itself into?

Exam Questions

Q1 Write a definition of a gene. [2 marks]

Q2 A gene has the following base sequence: AATGCAGGCTCT.
Write the base sequence of the mRNA molecule that's formed along this gene in the nucleus. [2 marks]

My genes are degenerate — there's a hole in the back pocket... *(I'll get my code)*

Quite a few terms to learn here — you're on the inescapable road to science geekville I'm afraid, and it's a road lined with crazy diagrams and strange words. The genetic code is sooo important — so make sure you understand what's going on. You need to learn the structure of mRNA and tRNA — it'll help you understand protein synthesis on the next page.

Protein Synthesis

You've learnt about the genetic code and types of RNA — now you can learn all about their role in making proteins. This stuff is biology at its most clever, and it's probably going on inside you right now. Weird.

*First Stage — **Transcription** Occurs in the **Nucleus***

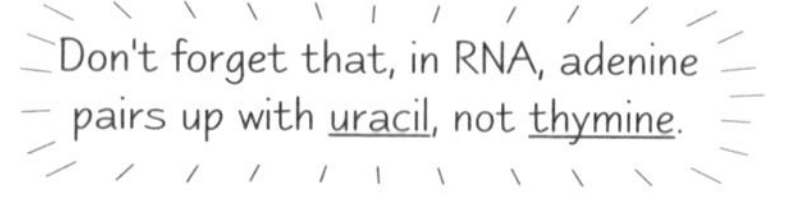

In **transcription** a **'negative copy'** of a **gene** is made. This copy is called **mRNA**.

1) A gene (a section of DNA) in the DNA molecule **uncoils** and the hydrogen bonds between the two strands in that section break, separating the strands.
2) One of the strands is then used as the **template** for transcription — it's called the **'sense strand'**.
3) Free **RNA nucleotides** in the nucleus line up alongside the template strand. Once the RNA nucleotides have **paired up** with their **complementary bases** on the DNA strand they're joined together by the enzyme **RNA polymerase**.
4) The strand that's formed is **mRNA**.
5) It then moves out of the nucleus through a nuclear pore, and attaches to a **ribosome** in the cytoplasm, where the next stage of protein synthesis takes place.
6) When enough mRNA has been produced, the uncoiled strands of DNA re-form the hydrogen bonds and **coil back into a double helix**, unaltered.

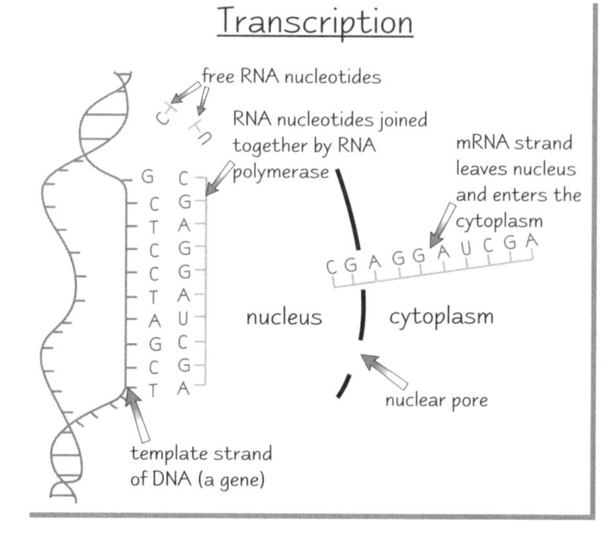

*Second Stage — **Translation** Occurs in a **Ribosome***

In **translation**, **amino acids** are stuck together to make a **protein**, following the order of amino acids coded for on the mRNA strand.

1) The **mRNA strand** has travelled to a ribosome in the cytoplasm, and attached itself.
2) All 20 **amino acids** needed to make human proteins are in the cytoplasm. tRNA molecules attach to the amino acids and transport them to the ribosome.
3) In the ribosome, a tRNA molecule binds to the start of the mRNA strand. This tRNA molecule has the **complementary anticodon** to the **first codon** on the mRNA strand, and attaches by **base pairing**. Then a second tRNA molecule attaches itself to the **next codon** on the mRNA strand in the **same way**.
4) The two amino acids attached to the tRNA molecules are joined together with a **peptide bond** (using ATP and an enzyme).
5) The first tRNA molecule then **moves away** from the ribosome, leaving its amino acid behind. The mRNA then **moves across** the ribosome by one codon and a third tRNA molecule binds to the **next codon** that enters the ribosome.
6) This process continues until there's a **stop codon** on the mRNA strand that doesn't code for any amino acid. You're left with a line of amino acids joined by peptide bonds. This is a **polypeptide chain** — the **primary structure** of a protein. The polypeptide chain moves away from the ribosome and translation is complete.

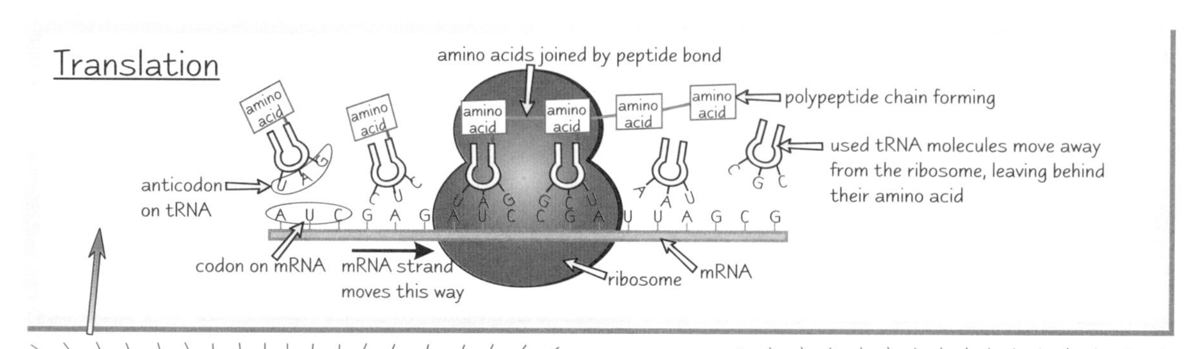

So, in protein synthesis, the sequence of codons on the mRNA strand determines the sequence of amino acids that makes up the primary structure of the protein.

When translation is complete, the polypeptide chain folds itself into its secondary and tertiary structure, and a protein is formed, e.g. an enzyme (see p.4).

Mutation

*Mutations happen naturally at random. But certain things (called **mutagenic agents**) make mutations happen more often. Mutagenic agents include tobacco smoke, some chemicals, radiation, ultraviolet light from the Sun and x-rays.*

Mutations Alter the *Base Sequence* of *DNA*

Mutations are changes in the base sequence of an organism's **DNA**. A gene codes for a particular protein, so if the sequence of bases in a gene changes, a **different protein** is produced. Mutations produce **new alleles** of genes. There are three main types of **gene mutation**:

1) **Addition** — extra base included, e.g.

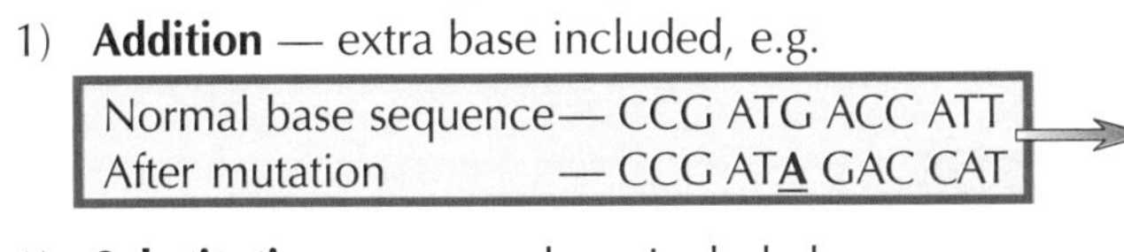

Normal base sequence— CCG ATG ACC ATT
After mutation — CCG AT**A** GAC CAT

Addition means the whole sequence of bases after the mutation is out by one base. All the codons after the mutation are different, and code for different amino acids. The resulting protein won't have the right 3D structure, so it won't work at all.

2) **Substitution** — wrong base included, e.g.

Normal base sequence— CCG ATG ACC ATT
After mutation — CC**T** ATG ACC ATT

Only one codon is affected by substitution. More than one codon can code for an amino acid, so if the new codon codes for the same amino acid as the old one, the sequence of amino acids will remain the same and the resulting protein won't change. But if the new codon codes for a different amino acid, a new, non-functioning protein will be formed.

3) **Deletion** — base missed out, e.g.

Normal base sequence— CCG AT**G** ACC ATT
After mutation — CCG ATA CCA TT

Deletion means the whole sequence of bases is out by one, just like with an addition mutation. This is also called a frame-shift, by the way.

Gene Mutations can Block *Metabolic Pathways*

Gene mutations can mean that the protein produced isn't the protein that the gene normally codes for. Remember that all **enzymes** are **proteins**.
If there's a mutation in the gene that codes for an enzyme, then that enzyme won't **fold up** properly, its **active site** will be the wrong shape and it **won't work**. Enzymes **catalyse** biological reactions, so **faulty enzymes** mean that the reaction they're supposed to catalyse **won't happen** properly.
When these reactions are vitally important for the metabolism of the cell, the mutation in a gene for an enzyme is **seriously bad news**.

Our metabolism is all the chemical reactions that take place in our cells to keep us alive. They usually happen in a series of small reactions, not one big one. The series of reactions are called metabolic pathways.

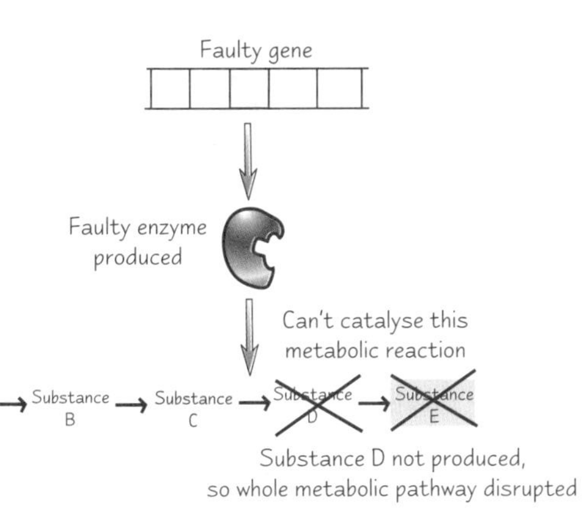

Practice Questions

Q1 What are the two main stages in protein synthesis?

Q2 Where does translation take place?

Q3 What is a mutation, and how can it affect protein structure?

Exam Question

Q1 Sickle cell anaemia is a genetic disease resulting from the following gene mutation:

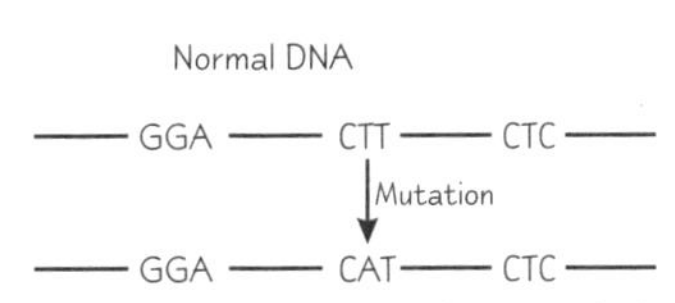

a) From information in the diagram, what type of mutation causes sickle cell anaemia? [1 mark]
b) Explain how this change in base has affected the protein formed. [3 marks]

mRNA codons join to tRNA anticodons?! — I need a translation please...

When you first go through protein synthesis it might make approximately no sense, but I promise its bark is worse than its bite. All those strange words disguise what is really quite a straightforward process — and the diagrams are dead handy for getting to grips with it. Keep drawing them yourself, 'til you can reproduce them perfectly.

Mitosis

I don't like cell division. There, I've said it. It's unfair of me, because if it wasn't for cell division I'd still only be one cell big. It's all those diagrams that look like worms nailed to bits of string that put me off.

Mitosis is Cell Division that Produces Genetically Identical Cells

Cells increase in number by **cell division**. **Mitosis** is a type of cell division where a parent cell divides to produce **two daughter cells** that are **genetically identical** to the parent cell. It's needed for the **growth** of multicellular organisms (like us) and for **repairing** damaged tissues. How else do you think we get from being a baby to being a big, strapping lass / lad — it's because the cells in our bodies divide and multiply. It's also used in **asexual reproduction**.

Chromosomes are Important for Cell Division

A **chromosome** is a **thread-like structure** made up of **one long molecule** of **DNA** that's **coiled** up very tightly so it's nice and **compact**. Chromosomes are found in the nucleus of plant and animal cells, and they play an important part in cell division. Mitosis produces cells that have the same number and type of chromosomes as the parent cell.

Mitosis has Four Division Stages plus Interphase

Mitosis is really one **continuous process**, but it's described as a series of **division stages** — prophase, metaphase, anaphase and telophase. **Interphase** comes before the division stages in the **cell cycle** — it's when cells grow and prepare to divide by replicating their DNA.

1) **Interphase** — The cell carries out normal functions, but also prepares to divide. It **replicates its DNA**, to double its genetic content. It also **replicates its organelles** so it has spare ones, and **increases its ATP content** (ATP provides the energy needed for cell division).

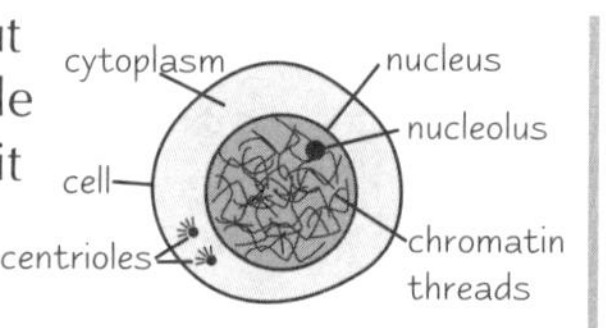

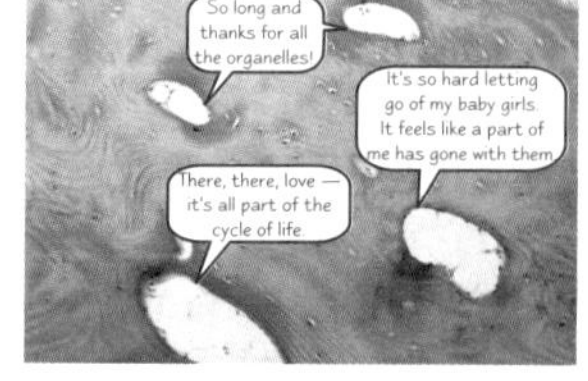

Mitosis can be a moving time.

2) **Prophase** — The chromosomes **condense**, getting shorter and fatter. Tiny bundles of protein called **centrioles** start moving to opposite ends of the cell, forming a network of protein fibres across it called the **spindle**. The nuclear membrane breaks down and chromosomes lie free in the cytoplasm.

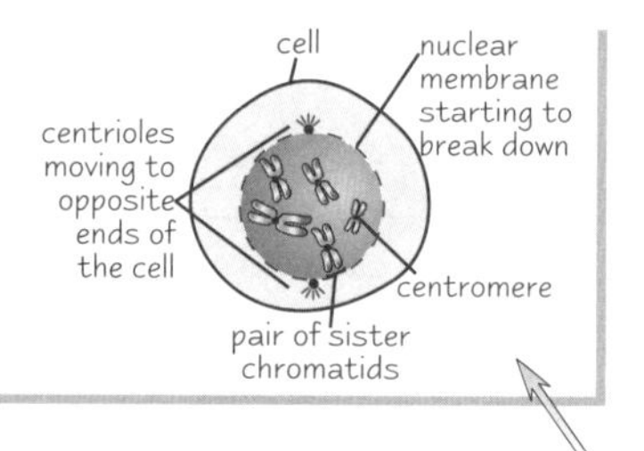

As mitosis begins, the chromosomes are made of two strands joined in the middle by a centromere. The separate strands are called chromatids. There are two strands because each chromosome has already made an identical copy of itself during interphase. When mitosis is over, the chromatids end up as one-strand chromosomes in the new daughter cells.

3) **Metaphase** — The chromosomes (each with two chromatids) line up along the middle of the cell and become attached to the spindle by their centromere.

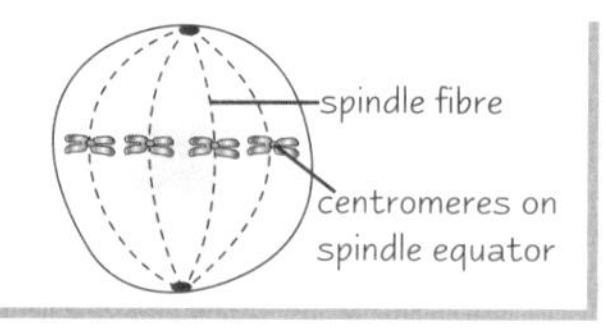

4) **Anaphase** — The centromeres attaching the chromatids to the spindles divide, separating each pair of sister chromatids. The spindles contract, pulling chromatids to opposite poles, centromere first.

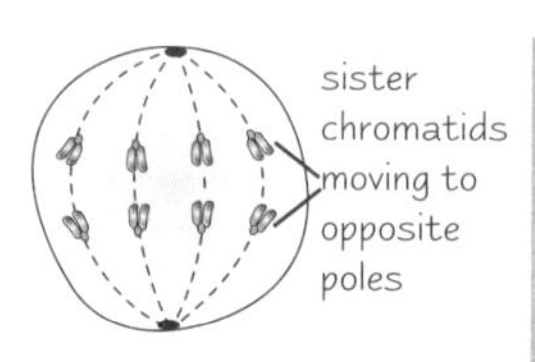

5) **Telophase** — The chromatids reach the **opposite poles** on the spindle. They uncoil and become long and thin again. They're now called **chromosomes** again.
A **nuclear membrane** forms around each group of chromosomes, so there are now **two nuclei**. The **cytoplasm divides** and there are now **two daughter cells** which are **identical** to the original cell. Mitosis is finished and each daughter cell starts the **interphase** part of the cell cycle to get ready for the next round of mitosis.

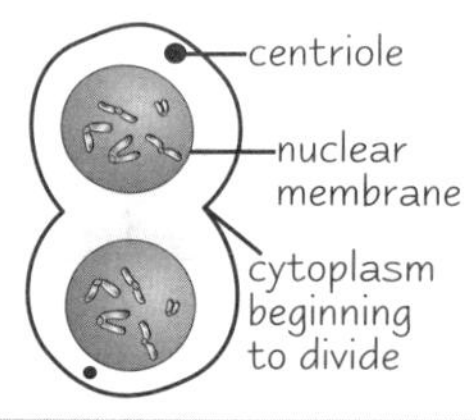

Cloning

Asexual Reproduction through Mitosis Produces Clones

Asexual reproduction needs only **one** parent. The offspring are normally **genetically identical** to this parent and are called **clones**. Asexual reproduction in plants can be natural or artificial:

1) **Tubers** grow **naturally** on plants like potatoes, and give clones when planted.
2) Some plant cells (called **meristems**) are able to divide and form all the cells needed to create a new plant. So humans can **artificially clone** plants by removing meristems and cultivating them in rooting powder and compost to create new plants, identical to the original. This is **vegetative propagation**.
3) Artificial cloning from tiny plant specimens (**explants**) is called **micropropagation**:

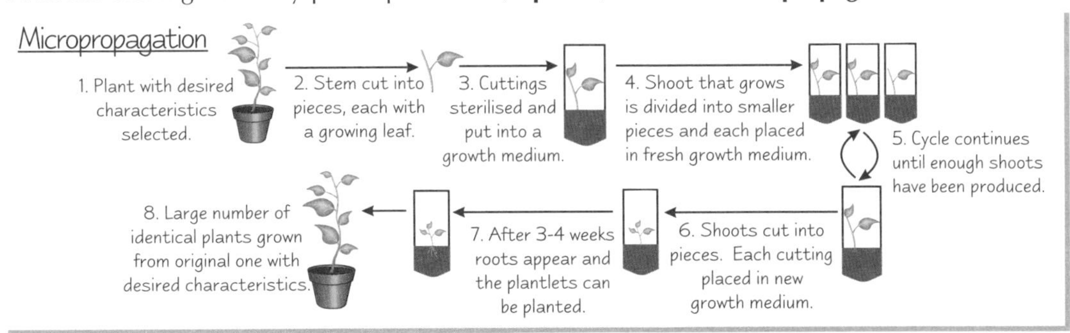

Cloning Animals is Possible but More Complicated

Here's one method of cloning animals with desirable features, like cows with **high milk yields**. Eggs from the best cow are fertilised in a petri dish with sperm from the best bull — this is called **in-vitro fertilisation**. The fertilised egg divides, giving a ball of genetically identical cells, which develop into an **embryo**. The young embryo can be **split** into separate cells — each cell grows into a new embryo that's genetically identical to the original one. These embryos are then transplanted into **surrogate** cows to develop.

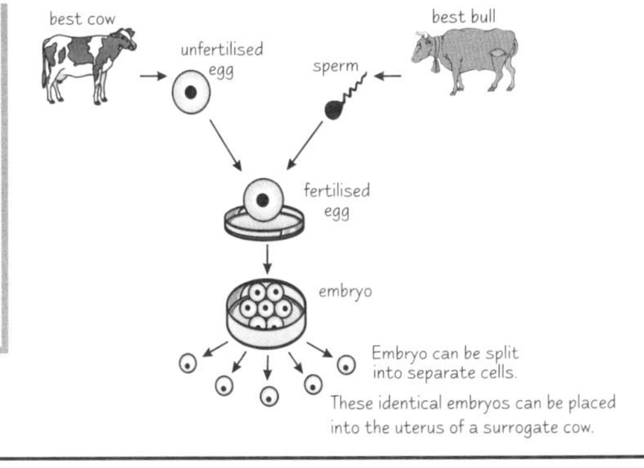

Practice Questions

Q1 What is a chromosome made up of?

Q2 What is meant by asexual reproduction?

Q3 Explain how animals with desirable characteristics can be cloned.

Exam Questions

Q1 The diagrams show cells at different stages of mitosis.

Cell A (labels Z, X, Y)

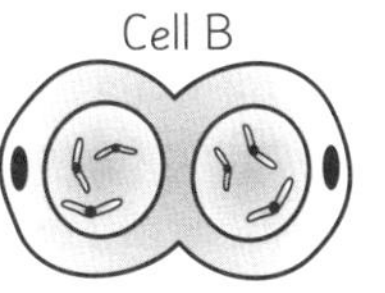

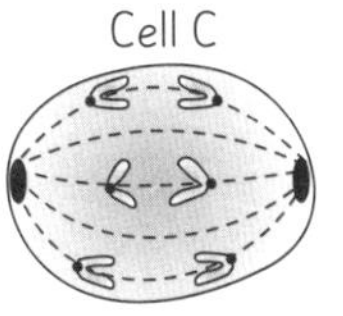

a) For each of the cells A, B and C, state the stage of mitosis, giving a reason for your answer. [6 marks]

b) Name the structures labelled X, Y and Z in cell A. [3 marks]

Q2 During which stages of the cell cycle would the following events take place?

a) DNA replication. [1 mark]

b) Formation of spindle fibres. [1 mark]

Doctor, I'm getting short and fat — don't worry, it's just a phase...

Quite a lot to learn in this topic — but it's all dead important stuff so no slacking. All cells undergo mitosis — it's how they multiply and how organisms like us grow and develop. Cloning is mad scientist territory — don't try it at home... just learn all those facts in case you ever want to clone a cow. Stranger things have happened, but I'm not sure where.

Gametes, Meiosis and Sexual Reproduction

*More cell division — lovely jubbly. Meiosis is the cell division used in sexual reproduction. It consists of two divisions, not one. The **second division** is exactly the same as mitosis, which is handy.*

DNA From One Generation is Passed to the Next by Gametes

Gametes are the **sperm** cells in males and the **ova** (egg cells) in females. They join together at **fertilisation** to form a **zygote**, which divides and develops into a **new organism**.

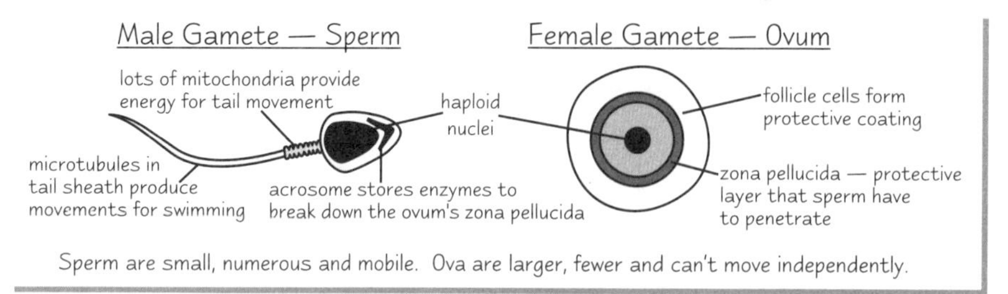

Sperm are small, numerous and mobile. Ova are larger, fewer and can't move independently.

Gametes have Half the Usual Number of Chromosomes

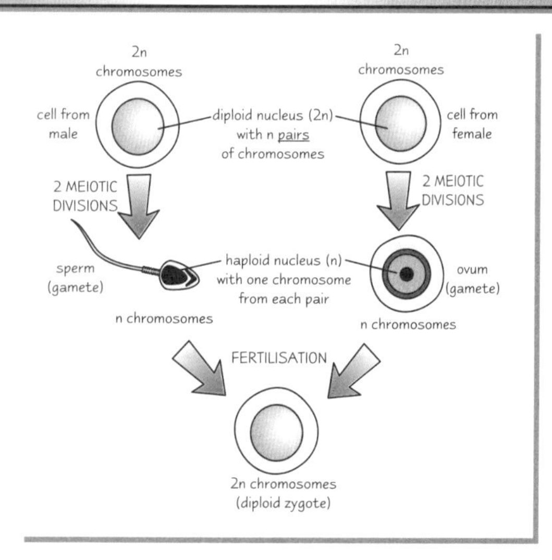

1) Normal **body cells** have the **diploid number** (**2n**) of chromosomes — meaning each cell contains **two** of each chromosome, one from the mum and one from the dad.
2) **Gametes** have a **haploid** (**n**) number of chromosomes — there's only one of each chromosome.
3) At **fertilisation**, a **haploid sperm** fuses with a **haploid egg**, making a cell with the normal diploid number of chromosomes. Half these chromosomes are from the father (the sperm) and half are from the mother (the egg).

Meiosis Halves the Chromosome Number

1) **Meiosis** is a type of cell division. It's essential for **sexual reproduction**. Cells that divide by meiosis are **diploid (2n)** to start with, but the cells that result from meiosis are **haploid (n)**. Without meiosis, you'd get **double** the number of chromosomes in each generation, when the gametes fused.
2) Meiosis happens in the **reproductive organs**. In humans, it's in the **testes** for males and the **ovaries** for females. In plants, it's in the **anthers** and **ovules**.
3) Unlike mitosis, there are **two divisions**. This **halves** the chromosome number.

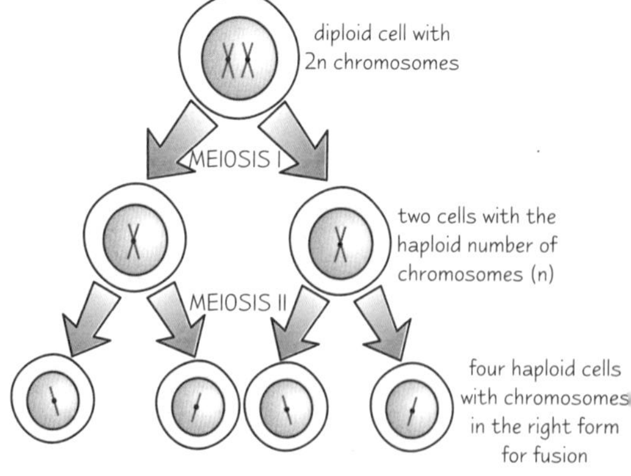

The two divisions in meiosis are called **meiosis I** and **II**.
Each division has 4 stages, just like mitosis — **prophase**, **metaphase**, **anaphase** and **telophase**.

1) In **Meiosis I** the **homologous pairs** of **chromosomes** are separated, which **halves** the number of chromosomes in the daughter cells.
2) **Meiosis II** is like mitosis — it separates the **pairs of chromatids** that make up each chromosome.
3) Unlike mitosis, which results in two genetically identical diploid cells, meiosis results in **four haploid cells** (gametes) that are **genetically different** from each other.

Gametes, Meiosis and Sexual Reproduction

A *Life Cycle* is the Sequence of Events from One Generation to the Next

In organisms which reproduce sexually, the cell life cycle is the period from the **fertilisation** of the cell to the fertilisation of its **daughter cells**.

Here's a diagram of the life cycle in human reproductive cells:

This is quite a common cell life cycle — but some organisms are different e.g. meiosis might take place at a different point in the cycle.

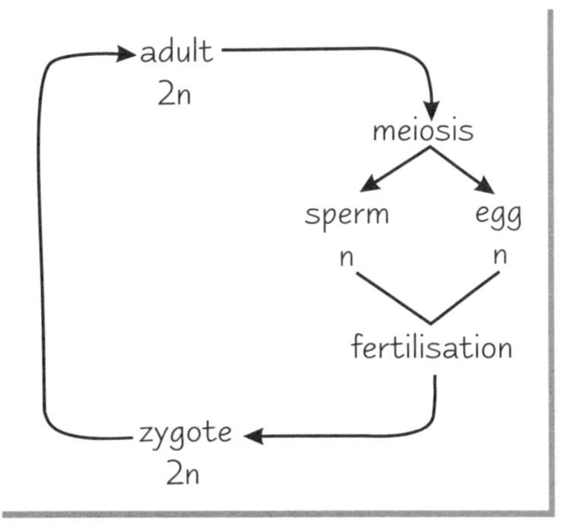

Exam questions might ask you to interpret a diagram of a life cycle — and point out where different events have happened. Remember:

1) When a diploid (2n) cell divides into two haploid (n) cells, then **meiosis** has taken place.
2) When two haploid (n) cells combine into a diploid (2n) cell, then **fertilisation** has taken place.
3) When a cell divides into two identical versions of itself, each with the same chromosome number, then **mitosis** has taken place.

Practice Questions

Q1 Explain what is meant by the terms "haploid" and "diploid".

Q2 Give two adaptations of sperm that allow them to perform their function.

Q3 How many divisions are there in meiosis?

Q4 What is the life cycle of a sexually reproducing organism's cells?

Exam Question

Q1 The diagram below shows stages of meiosis in a human ovary. Each circle represents a cell.

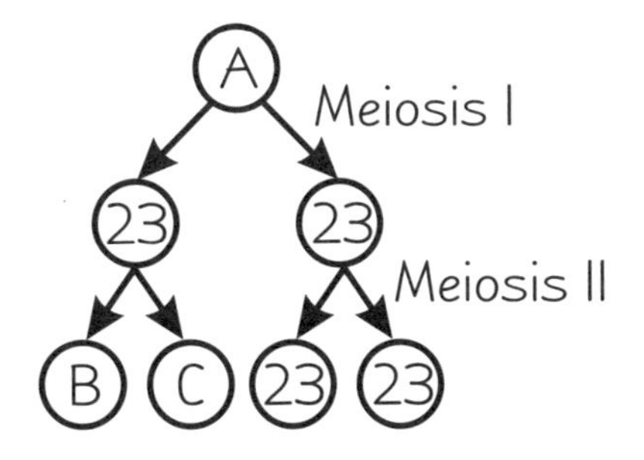

a) How many chromosomes would be found in cells A, B and C? [3 marks]

b) Explain why it's important for gametes to have half the number of chromosomes as normal body cells. [2 marks]

Reproduction isn't as exciting as some people would have you believe...

For some reason, this stuff can take a while to go in (insert own joke). But that's no excuse to just sit there staring frantically at the page and muttering "I don't get it," over and over again. Use the diagrams to help you understand — they look evil, but they really help. The key thing is to understand what happens to the ***number of chromosomes*** *in meiosis.*

Genetic Engineering

Genetic engineering is a dead popular exam topic because it shows how biology relates to real life — and examiners love all that stuff. These pages explain the process for manufacturing genes and, you've got to admit, it's kinda cool.

Genetic Engineering has **Important Medical Uses**

Sometimes humans can't produce a certain **protein** because the **gene** that codes for it is **faulty**.
By creating recombinant DNA (see below), genes can be artificially **manufactured** to replace the faulty genes.
This has important **medical benefits**. For example:

1) **Diabetics** don't have the healthy gene needed to make the protein **insulin**, which controls blood sugar levels.
2) **Haemophiliacs** lack the healthy gene that codes for the protein **factor VIII**, which allows blood to clot.
3) Genetic engineering allows these genes to be manufactured in bacterial cells and used to help sufferers.

Recombinant DNA is like **'Home-Made'** DNA

DNA has the **same** structure of **nucleotides** in **all organisms**. This means you can **join together** a piece of DNA from one organism and a piece of DNA from another organism.

DNA that has been **genetically engineered** to contain DNA from another organism is called **recombinant DNA**. It has **useful applications**. In the example below, a **human gene** coding for a **useful protein** is inserted into a bacterium's DNA. When the **bacterium reproduces**, the **gene** is **reproduced** too. The gene is **expressed** in the bacterium, and so the bacterium **produces** the **protein** which is coded for by the gene.

1) First you **find** the gene you want in the donor cell. The **DNA** containing the gene is **removed** from the cell and any **proteins** surrounding the extracted DNA are removed using **peptidase enzymes**.

2) Next you **cut out** the **useful gene** from the DNA using **restriction endonuclease** enzymes. These leave a **sticky end** (tail of unpaired bases) at each end of the useful gene. This stage is called **restriction**.

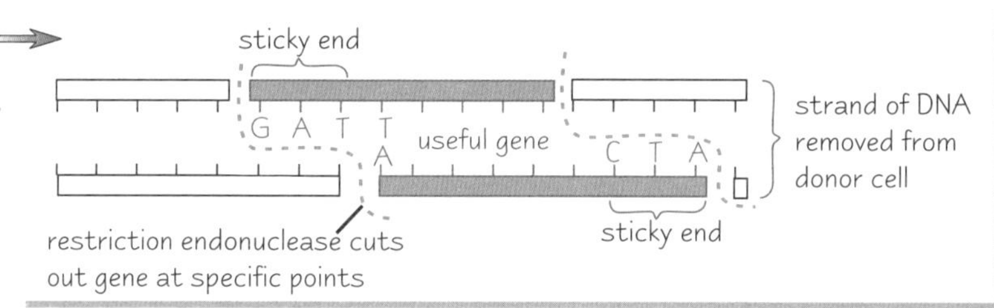

3) Then you need to **prepare** the other bit of DNA that you're **joining** the useful gene to (called a **vector**). The main vectors used are **plasmids** — these are small, **circular molecules** of DNA found in **bacteria**. They're useful as vectors because they can **replicate** without interfering with the bacterium's own DNA. The same **restriction enzyme** is used to **cut out** a section of the plasmid. The **sticky ends** that are left have bases that are **complementary** to the bases on the sticky ends of the **useful gene**.

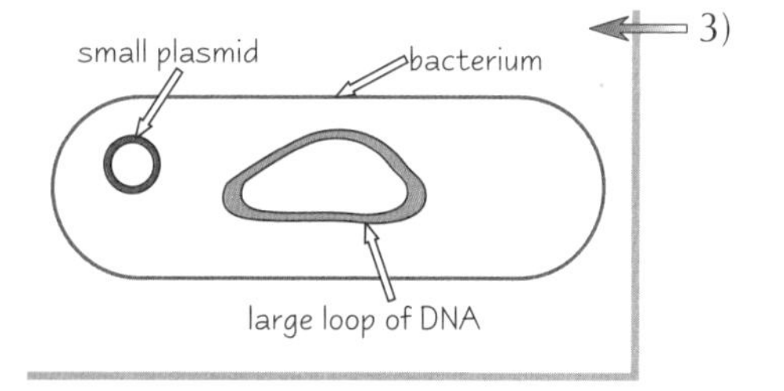

4) Finally, you need to **join** the **useful gene** to the **plasmid vector DNA** — this process is called **ligation**. This is where the **sticky ends** come in — **hydrogen bonds** form between the **complementary bases** of the sticky ends. The DNA molecules are then 'tied' together with the enzyme **ligase**. The new DNA is called **recombinant DNA**.

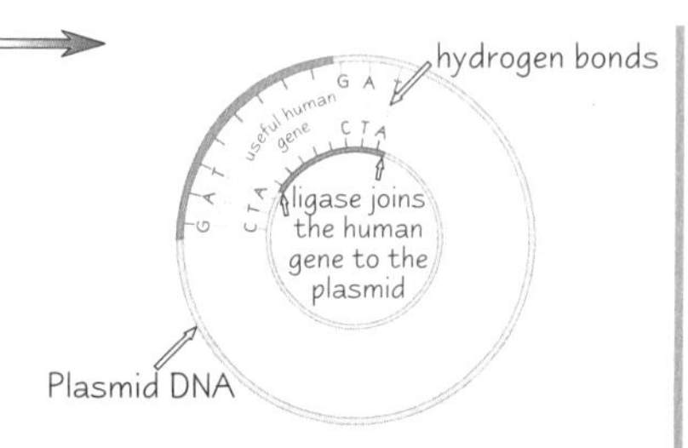

The useful gene is now in the recombinant DNA of a plasmid vector. This version can then be inserted into a bacterium. Then the bacterium is made to reproduce loads, so you get loads of copies of the gene. Clever, huh. p.44 has more on this.

Genetic Engineering

Reverse Transcriptase makes DNA from mRNA

Transcription is when mRNA is made from DNA (see p.36). **Reverse transcriptase** is a nifty enzyme that runs this process backwards and makes DNA from mRNA. It's useful in genetic engineering because **cells** that make **specific proteins** usually contain **more mRNA molecules than genes**. For example, **pancreatic cells** produce the protein **insulin**. They have loads of mRNA molecules on their way to ribosomes to make the insulin — but only two copies of the DNA gene for insulin, one on each pair of chromosomes.

Here's how reverse transcriptase can help make use of all this mRNA:

1) **mRNA** is **extracted** from donor cells.
2) The mRNA is mixed with **free DNA nucleotides** and **reverse transcriptase**. The reverse transcriptase uses the mRNA as a **template** to synthesise a new strand of **complementary DNA**.
3) The complementary DNA can be made **double-stranded** by mixing it with DNA nucleotides and **polymerase enzymes**. Then the useful gene from the double-stranded DNA is inserted into a plasmid so the bacteria can make lots of the product of the gene.

Practice Questions

Q1 What is recombinant DNA?

Q2 What do restriction endonuclease enzymes do?

Q3 How are useful genes joined to plasmids?

Q4 How does reverse transcriptase work?

Q5 Identify the people in these photographs. Have a cup of tea.

Exam Questions

Q1 Plant breeders have found a variety of cabbage that is resistant to a pest called root fly. This is because the cabbage produces a protein that inhibits one of the fly's digestive enzymes. Plant breeders want to manufacture this protein so they can insert it into carrots too. Describe how scientists could:

a) remove the gene that produces the inhibitor from the cabbage. [2 marks]

b) insert this gene into the DNA of a bacterium. [2 marks]

Q2 One technique used to produce human factor VIII by genetic engineering involves inserting a gene for human factor VIII into the DNA of a bacterium. Name the enzyme which would be used to:

a) cut the bacterial DNA. [1 mark]

b) insert the DNA for human insulin into the cut bacterial DNA. [1 mark]

Monkey vectors — Plasmid of the Apes...

You see, biology isn't just an evil conspiracy to keep students busy — it has loads of really important uses in real life. For example, sufferers of genetic diseases now have a far greater chance of having successful treatment, which is nice. Round of applause for biology, that's what I say.

Genetic Engineering

Page 42 explained how a useful gene can be 'cut out' of human DNA and joined with plasmid DNA. This genetically engineered plasmid DNA then needs to be inserted into bacteria for TWO HIGHLY EXCITING REASONS —
1) so when the bacteria reproduce asexually, all the new 'offspring' bacteria contain copies of the useful gene,
2) so the gene is expressed in the bacteria and the bacteria start to make the protein coded for by the gene.

Genes are Moved into Micro-organisms using Vectors

Vectors carry recombinant DNA containing useful genes into microbes. They're usually plasmids or viruses, like the **λ** (**Lambda**) **phage virus**.

1) **Host bacteria** have to be **persuaded** to take up **plasmid** vectors. They're placed into cold **calcium chloride** solution to make their cell membranes more **permeable**. Then the plasmids are added and the mixture warmed up. The **sudden cold** causes some of the bacteria to take up the plasmids.
2) When a **phage** is used as a vector, the phage DNA is combined with the useful gene to make recombinant DNA. The phage then **infects** a bacterium by injecting its DNA strand into it. The phage DNA is then **integrated** into the bacterium's DNA.
3) Bacteria that have taken up vectors are said to be **transformed**.

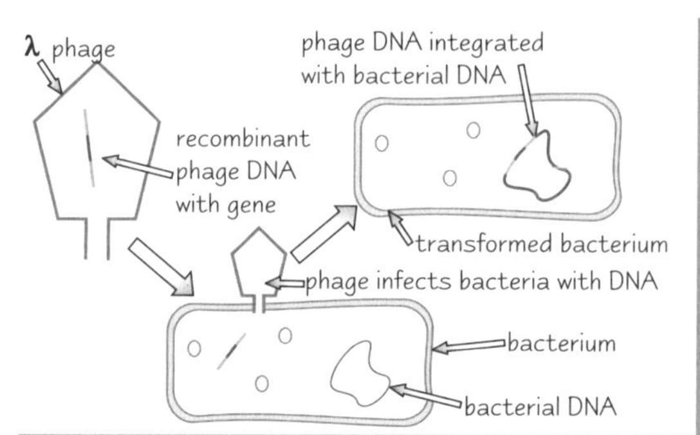

Transformed Bacteria can be Identified Using Marker Genes

Not all the bacteria will take up vectors, so you need a way to **pick out** the **transformed** bacteria. The easiest way is to use **antibiotic marker genes**.

1) When recombinant DNA is produced in plasmids, a marker gene for **antibiotic resistance** is inserted into the plasmid as well as the donor gene. This means bacteria that have been transformed contain both the donor gene **and** the gene for antibiotic resistance.
2) After being mixed with plasmids, the bacteria are cultured on an agar plate called the **master plate**.
3) Once bacteria have grown on the master plate, **replica plating** is used to isolate **transformed bacteria**:

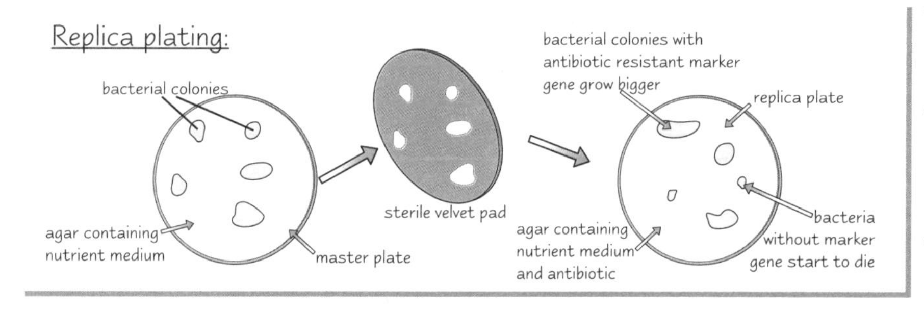

- A **sterile velvet pad** is pressed onto the master plate. This picks up some bacteria from each colony.
- The pad is pressed onto a fresh agar plate, **containing an antibiotic**. Some of the bacteria from each colony are transferred onto the agar surface.
- Only transformed bacteria can grow and reproduce on the replica plate — the others don't contain the antibiotic-resistant gene, so they stop growing.

Industrial Fermenters are Used for Large Scale Protein Production

Once a useful gene is in a bacterium, it has to start working to produce the **protein**.
An **industrial fermenter** is used to culture the bacteria and produce a **large amount** of the gene product (the protein).

1) A **promoter** gene is often included along with the **donor** gene when the **recombinant DNA** is made. This '**switches on**' the useful donor gene so it starts making the **protein**.
2) The **bacteria** containing the recombinant DNA are grown or cultured in a **fermenter**. Inside the fermenter, they're given the **ideal conditions** needed for rapid growth. They **reproduce quickly**, until there are millions of bacteria inside the fermenter.
3) The **plasmids**, including plasmids made of recombinant DNA, **replicate** at each cell division — so each new bacterium contains the useful gene.
4) As the bacteria grow, they start producing the **human protein** that the donor gene in the plasmid codes for, e.g. **human factor VIII** or **human insulin**.
5) The bacteria can't use human protein, so it **builds up** in the medium inside the fermenter. When enough has built up, it can be **extracted** and processed for use.

Genetic Engineering

Genetically Engineered Microbes are **Micro-factories**

Genetically engineered microbes can be used to produce large quantities of many **useful substances**. E.g.

1) **Antibiotics** like penicillin.
2) **Hormones**, like insulin and human growth hormone.
3) **Enzymes** used in manufacturing processes, like those in biological washing powders.

Genetic Engineering Raises **Ethical** and **Moral** Issues

Genetic engineering is the subject of an awful lot of arguments, fist fights and boring think-pieces in the Guardian. Here are some of the **ethical** and **moral issues** which need to be considered, before anyone says the dreaded words, "For my first child, I want a girl who looks like Victoria Beckham (but with Meg Ryan's hair)."

Potential Benefits of Genetic Engineering	Concerns over Genetic Engineering
Specific medicines could be developed to treat diseases. A current example of this is the production of genetically engineered human insulin to control diabetes.	Risk of foreign genes entering **non-target** organisms and disrupting functions.
Faulty genes could be identified and replaced, preventing genetically inherited diseases (see p.46).	Risk of **accidental transfer** of unwanted genes, which could damage the recipient.
Parents could make sure their babies didn't have faulty genes before they were born.	Doctors might be under pressure to implant embryos that could provide **transplant material for siblings**. Those who can afford it might decide which characteristics they wish their children to have (**designer babies**), creating a 'genetic underclass'. The **evolutionary consequences** of genetic engineering are unknown. **Religious concerns** about 'playing God'.
Crops could be engineered to give **increased yields** or exploit new habitats	**Health concerns** over human consumption of genetically modified foods.

Practice Questions

Q1 Explain how recombinant DNA is put into bacteria using vectors.

Q2 What is a marker gene?

Q3 How can you use replica plating to isolate transformed bacteria?

Q4 Briefly describe the process of large-scale protein production, using bacteria containing recombinant DNA.

Q5 Give three examples of useful substances which can be produced by genetically engineered microbes.

Exam Question

Q1 A human gene was combined with a plasmid which also contained a gene coding for resistance to ampicillin (an antibiotic). The plasmid was added to a bacterial culture.

a) Which technique could be used to find out whether the bacteria had taken up the plasmid containing the gene? [1 mark]

b) Explain how this technique works. [5 marks]

Transforming bacteria — all they need is a spot of paint...

The antibiotic-resistance gene is the genetic marker that you need to know about — learn how it's used to identify transformed bacteria, including how replica plating isolates the transformed bacteria. And then... and then... maybe just lay your head down on the desk... and dream sweet dreams. Nah. Not really. Do more revision. :)

Gene Therapy

Gene therapy isn't about group hugs and biscuits — it's to do with replacing faulty genes with healthy ones.

Gene Therapy Replaces Faulty Genes with Healthy Ones

Gene therapy replaces defective genes with healthy ones to treat **genetically inherited diseases**. The **cause** of the disease is tackled, not just the symptoms. The **healthy gene** is isolated and **cloned** and inserted into the cells where it's needed. The healthy gene **codes** for the **protein** that the **sufferer** of the disease **lacks**.
Cystic fibrosis is a genetic disease that can be treated by gene therapy.

Cystic Fibrosis is Caused by a Defect in the CFTR Gene

People who suffer from cystic fibrosis produce **thick, sticky mucus**. This stays in the lungs, causing them to become congested, which makes breathing difficult. The sticky mucus is caused by a **mutation** in the **CFTR gene**.

1) The **CFTR protein** is found in the cell membranes of **epithelial cells** lining the **airways** and the **gut**. It's a **channel protein**, which lets **chloride ions** out of the cells.
2) The **defective** CFTR protein is missing an **amino acid**. This affects the **3D shape** of the protein so that chloride ions can't move out of the cell.
3) The **ion concentration** in the cell increases, so more water's retained instead of moving out by osmosis. The **mucus lining** the airways becomes **thicker** and **stickier** as a result.

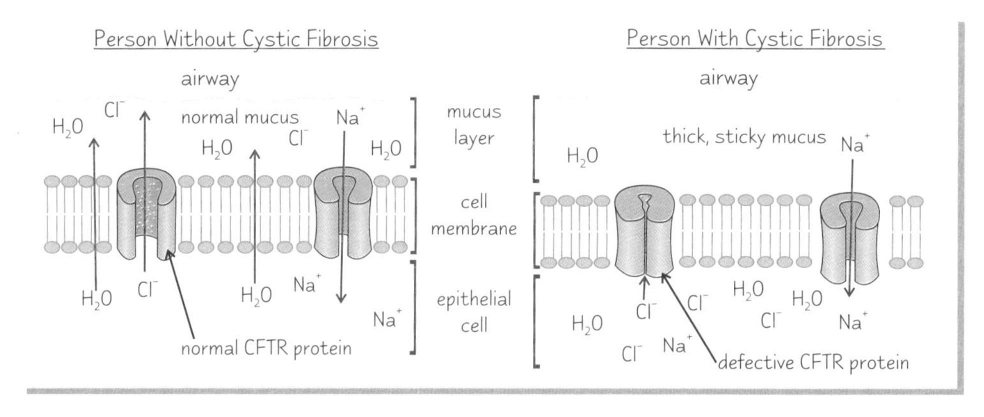

Vectors Transfer Healthy CFTR Genes in Epithelial Cells

Two main types of **vectors** are used to transfer healthy CFTR genes into body cells.

1) **Liposome vectors**. Recombinant DNA technology is used to put healthy CFTR genes into plasmids which are 'wrapped' in **liposomes**. Liposomes are minute lipid droplets that fuse with the cell membrane. The liposomes are sprayed into the lungs using **aerosols**. Liposomes can be produced on a large scale and patients don't develop an **immune reaction** to them.

2) **Adenovirus vectors**. Adenoviruses are the viruses that cause the **common cold**. First they're **inactivated**, by disabling the genes needed for them to replicate. Then the healthy CFTR gene is combined with the virus DNA. Then the viruses are **sprayed** into the lungs, where they **inject** the recombinant DNA into cells (see p.44).

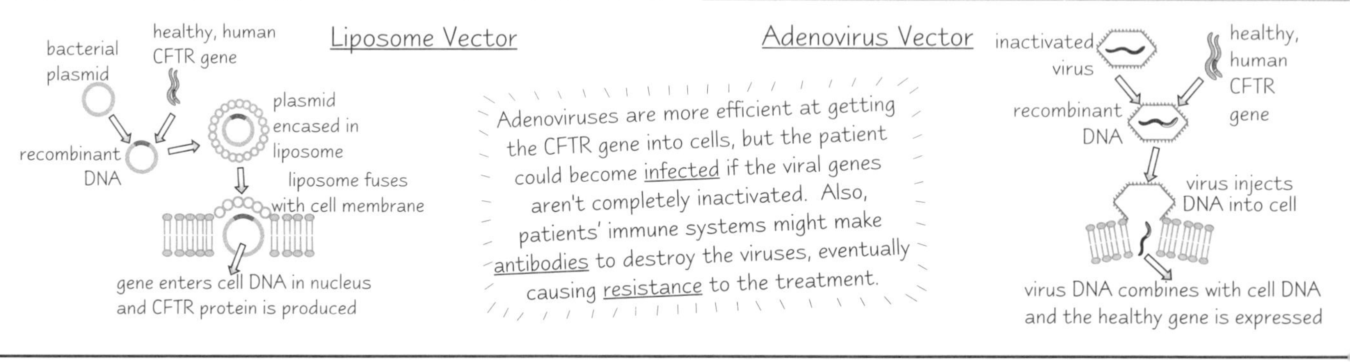

Gene Therapy

Important **Human Proteins** *are Produced by* **Genetically Engineered Animals**

People who can't produce the **glycoprotein alpha-1-antitrypsin (AAT)** suffer from **emphysema**.
Sheep can be genetically engineered to produce human **AAT**, which they secrete in their milk, bless 'em.
This can then be used to treat emphysema, and it can also help people with cystic fibrosis.

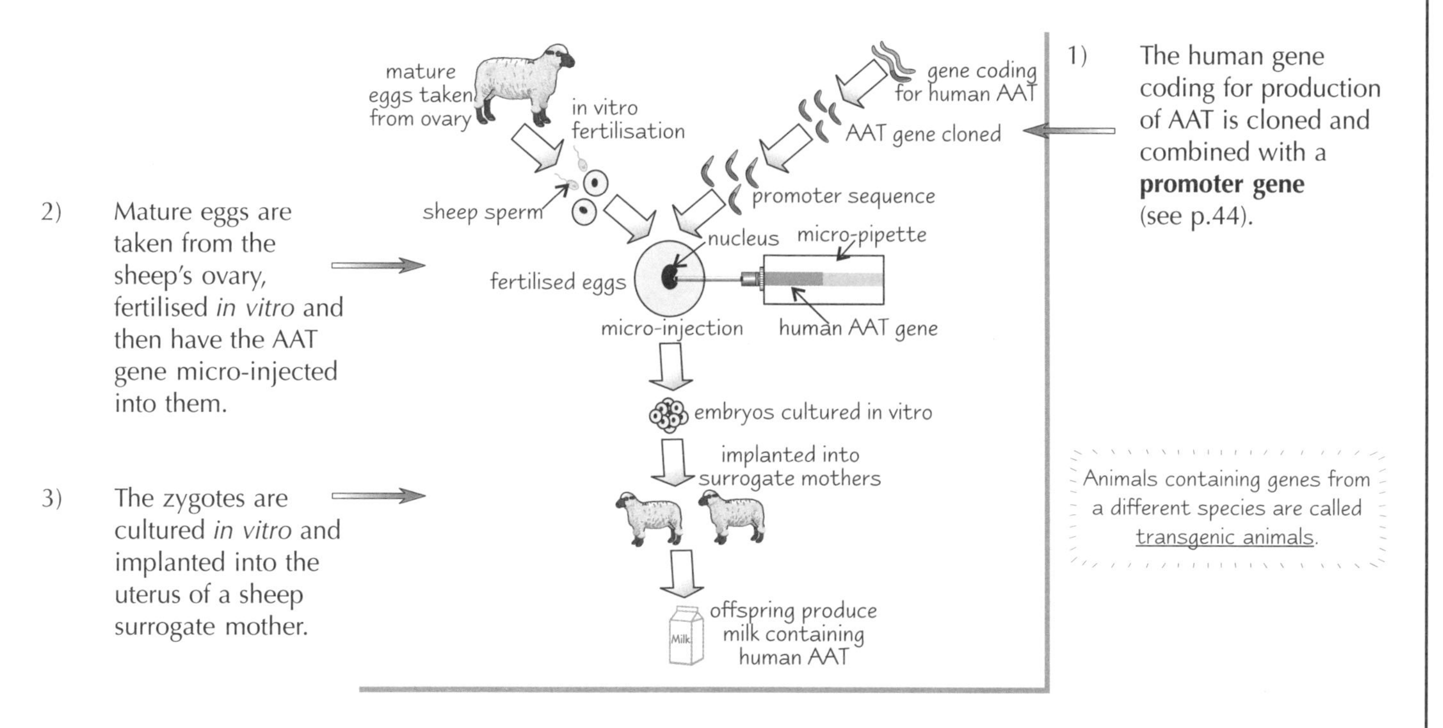

1) The human gene coding for production of AAT is cloned and combined with a **promoter gene** (see p.44).

2) Mature eggs are taken from the sheep's ovary, fertilised *in vitro* and then have the AAT gene micro-injected into them.

3) The zygotes are cultured *in vitro* and implanted into the uterus of a sheep surrogate mother.

Animals containing genes from a different species are called transgenic animals.

Practice Questions

Q1 What is gene therapy?

Q2 What are the symptoms of cystic fibrosis?

Q3 Give two examples of how humans can benefit from genetically modified organisms.

Q4 Which two vectors can be used to transfer healthy CFTR genes into human cells?

Exam Questions

Q1 a) Explain why there is a decrease in the movement of chloride ions and water in the airway epithelial cells of someone with cystic fibrosis. [4 marks]

b) Describe one way that the normal gene could be placed into the lung cells of someone suffering from cystic fibrosis. [5 marks]

Q2 People who suffer from haemophilia lack a clotting agent, factor VIII, in their blood.
They can be treated with factor VIII extracted from the milk of genetically engineered sheep.

a) Describe how sheep can be genetically engineered to produce human factor VIII in their milk. [3 marks]

b) Give one advantage of using genetically engineered factor VIII instead of factor VIII that has been extracted from blood donations. [1 mark]

What do you get if you cross Aladdin with a cow? — a transgenie...

Gene therapy is good news for cystic fibrosis sufferers — you need to learn what causes the disease and how it's treated by replacing the faulty gene. Make sure you know the difference between liposome and adenovirus vectors. Then learn about how we get important proteins from inserting the right genes into animals. Mmmm, milky.

Genetic Fingerprinting

The tiniest bit of DNA is enough to identify you. You're shedding hairs and flakes of skin-scum all the time (nice) — it's like leaving a personal trail of, well, ***you*** *wherever you go. Scary (and quite grim) thought.*

The **Polymerase Chain Reaction** (PCR) Creates Millions of **Copies** of **DNA**

Some samples of DNA are too small to analyse. The **Polymerase Chain Reaction** makes millions of copies of the smallest sample of DNA in a few hours. This **amplifies** DNA, so analysis can be done on it. PCR has **several stages**:

1) The DNA sample is **heated** at **95°C**. This breaks the hydrogen bonds between the bases on each strand. But it **doesn't** break the bonds between the ribose of one nucleotide and the phosphate on the next — so the DNA molecule is broken into **separate strands** but doesn't completely fall apart.

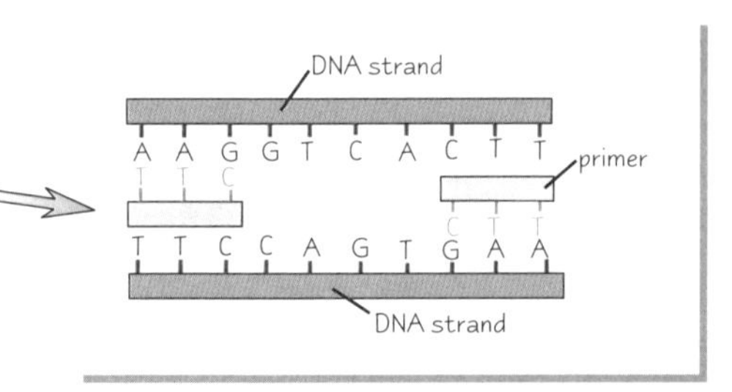

2) **Primers** (short pieces of RNA) are attached to both strands of the DNA — these will tell the **enzyme** where to **start copying** later in the process. They also stop the two DNA strands from joining together again.

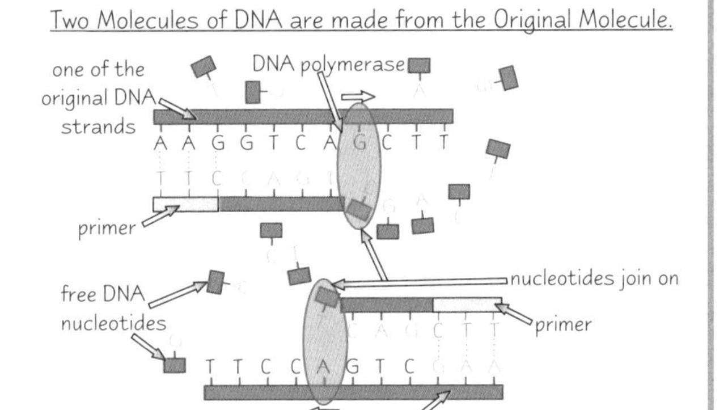

3) The DNA and primer mixture are **cooled** to **40°C** so that the primers can **fully bind on** to the DNA.

4) Free **DNA nucleotides** and the enzyme **DNA polymerase** are added to the reaction mixture. The mixture is heated to **70°C**. Each of the original DNA strands is used as a **template**. Free DNA nucleotides pair with their complementary bases on the template strands. The DNA polymerase attaches the new nucleotides together into a strand, starting at the primers.

5) The cycle starts again, using **both** molecules of DNA. Each cycle **doubles** the amount of DNA.

You can **Identify People** from their **DNA** by **Cutting** it into **Fragments**

It's possible to **identify a person** from a sample of their DNA, if the sample is big enough. This is done by using **enzymes** to cut the DNA up into **fragments**, then looking at the **pattern** of fragments, which is **different** for everyone (except identical twins).

1) To **cut up** the DNA into DNA fragments you add specific **restriction endonuclease** enzymes to the DNA sample — each one **cuts** the DNA every time a **specific base sequence** occurs. **Where** these base sequences occur on the DNA **varies** between everyone (except identical twins), so the number and length of DNA fragments will be different for everyone.

2) Next you use the process of electrophoresis to separate out the DNA fragments by size:

How Electrophoresis Works:

1) The DNA fragments are put into **wells** in a slab of **gel**. The gel is covered in a **buffer solution** that **conducts electricity**.
2) An **electrical current** is passed through the gel. DNA fragments are **negatively charged**, so they move towards the positive electrode. **Small** fragments move **faster** than large ones, so they **travel furthest** through the gel.
3) By the time the current is switched off, all the fragments of DNA are **well separated**.

Electrophoresis

DNA moves towards the anode, with smallest fragments moving furthest

−ve cathode

wells

DNA fragment (invisible)

gel, with buffer solution on top

+ve anode

In electrophoresis the DNA fragments **aren't visible** to the eye — you have to do something else to them before you can **see their pattern**. Coincidentally, that's what the next page is all about...

Genetic Fingerprinting

Gene Probes Make the Invisible 'Genetic Fingerprint' Visible

DNA fragments separated by electrophoresis are invisible.
A radioactive DNA probe (also called **gene probe**) is used to show them up:

1) A **nylon membrane** is placed over the electrophoresis gel, and the DNA fragments **bind** to it.
2) The DNA fragments on the nylon membrane are **heated** to separate them into **single strands**.
3) **Radioactive gene probes** are then put onto the nylon membrane. (It's the **phosphorus** in the gene probes' sugar-phosphate backbones that's radioactive.) The probes are warmed and **incubated** for a while so that they'll attach to any bits of complementary DNA in the DNA fragments.

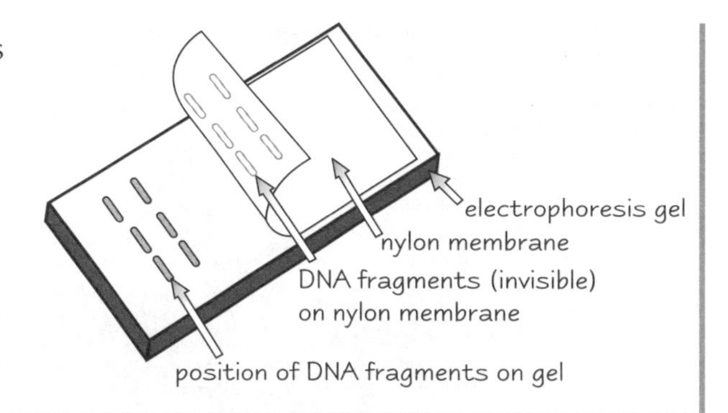

4) The nylon membrane is then put on top of unexposed **photographic film**. The film goes **dark** where the radioactive gene probes are **present**, which **reveals the position** of the **DNA fragments**. The pattern is different for every human — it's **unique** like a fingerprint.

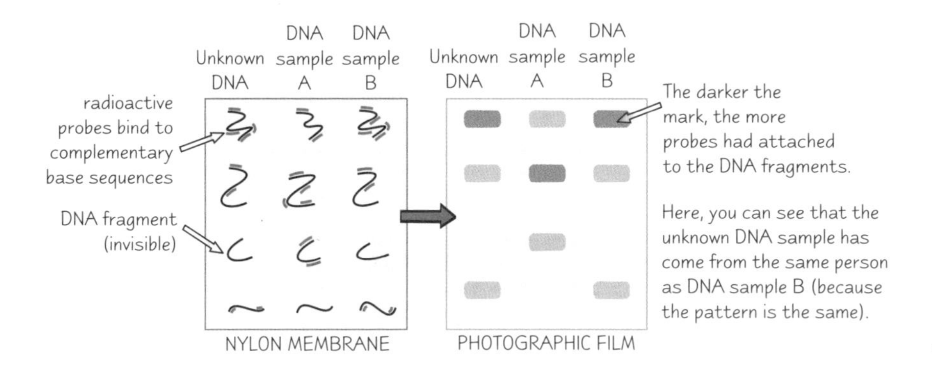

Genetic fingerprinting is incredibly useful. **Forensic investigations** use it to confirm the identity of suspects from blood, hair, skin, sweat or semen samples left at a crime scene, or to establish the identity of victims. **Medical investigations** use the same technique for **tissue typing, paternity tests** and **infection diagnosis.**

Practice Questions

Q1 Give the full name of the technique used to increase the amount of DNA from a very small sample.

Q2 Name the enzyme added to DNA before gel electrophoresis.

Q3 Which part of a gene probe is radioactive?

Q4 State three uses of genetic fingerprinting in medical investigations.

Q5 Name the enzyme used in PCR.

Exam Questions

Q1 Police have found incriminating DNA samples at the scene of a murder.
They have a suspect in mind, and want to prove that the suspect is guilty.

a) Name the technique that the police could use to confirm the guilt of the suspect. [1 mark]

b) Explain how this technique would be carried out. [5 marks]

You see that sweat stain there — that's you, that is...

Genetic fingerprinting has revolutionised medicine and forensic science. Remember that the PCR amplifies small DNA samples so genetic analysis can be done. You need to learn the PCR process, plus the process of electrophoresis and how it's involved in genetic fingerprinting. Also, learn what genetic fingerprinting is used for in the real world.

The Mammalian Heart

*Blood pumps **continually** round your body. It's happening right now as you read this rather long and dull sentence that I'm writing to keep you reading without stopping to try and highlight the **unceasingness** of blood flow. And relax.*

The Heart Consists of Two Muscular Pumps

The diagram below shows the **internal structure** of the heart. The **right side** of the heart pumps **deoxygenated blood** to the **lungs** and the **left side** pumps **oxygenated blood** to the **whole body**. NB — the **left and right side** are **reversed** on the diagram, 'cos it's the left and right of the person that the heart belongs to.

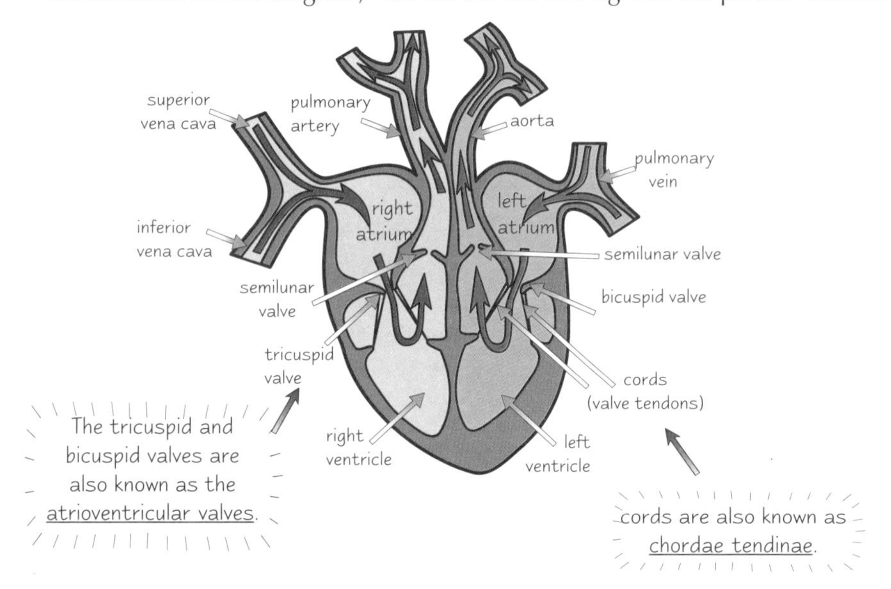

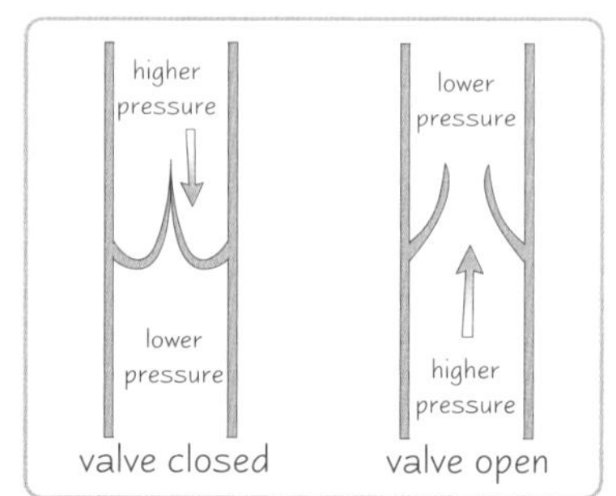

The **valves** only open one way — whether they open or close depends on the relative pressure of the heart chambers. If there's higher pressure behind a valve, it's forced open, but if pressure is higher above the valve it's forced shut.

Each Bit of the Heart is Adapted to do its Job Effectively

1) The **left ventricle** of the heart has thicker, more muscular walls than the **right ventricle**, because it needs to contract powerfully to pump blood all the way round the body. The right side only needs to get blood to the lungs, which are nearby.
2) The **ventricles** have thicker walls than the **atria**, because they have to push blood out of the heart whereas the atria just need to push blood a short distance into the ventricles.
3) The **tricuspid** and **bicuspid** valves link the atria to the ventricles and stop blood getting back into the atria when the ventricles contract.
4) The **semilunar valves** stop blood flowing back into the heart after the ventricles contract.
5) The **cords** attach the atrioventricular valves to the ventricles to stop them being forced up into the atria when the ventricles contract.

The Cardiac Cycle is Often Shown using a Graph

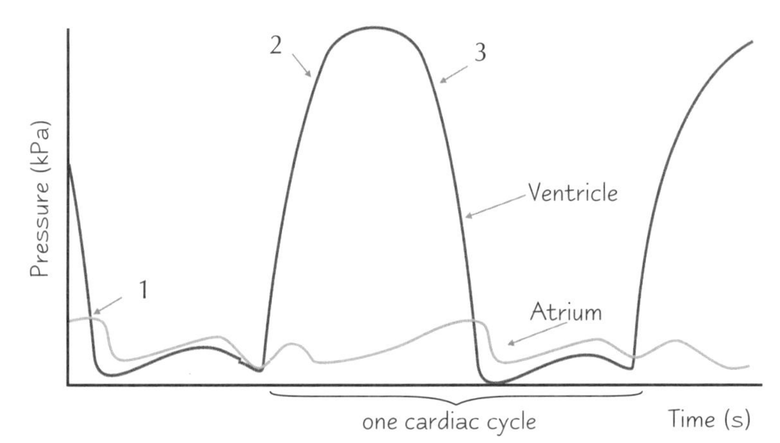

The graph shows the changes in pressure as the heart chambers contract and relax. The lines represent the **ventricle** and the **atrium**.

The next page describes what's happening at each stage of the graph.

Make sure that you're familiar with this graph — it comes up pretty often in exams.
It's better to learn it now than spending ages trying to get your head round it in the exam.

The Mammalian Heart

*The **Cardiac Cycle** Pumps Blood Round the Body*

The cardiac cycle is an ongoing sequence of **systole** (contraction) and **diastole** (relaxation) of the atria and ventricles that keeps blood continuously circulating round the body. The systole and diastole alter the **volume** of the different heart chambers, which alters **pressure** inside the chambers. This causes **valves** to open and close, which directs the **blood flow** through the system.
There are 3 stages:

1 Ventricular diastole, atrial systole

The **ventricles both relax**. The atria then contract, which decreases their volume. The resultant higher pressure in the atria causes the atrioventricular valves to open. This forces blood through the valves into the ventricles (point 1 on the cardiac graph on p.50).

2 Ventricular systole, atrial diastole

The **atria relax** and the **ventricles then contract**. This means pressure is higher in the ventricles than the atria, which shuts off the atrioventricular valves to prevent backflow. Meanwhile, the high pressure opens the semilunar valves and blood is forced out into the pulmonary artery and aorta (point 2 on the cardiac graph).

3 Ventricular diastole, atrial diastole

The **ventricles and the atria both relax**, which increases volume and lowers pressure in the heart chambers. The higher pressure in the pulmonary artery and aorta closes the semilunar valves to prevent backflow (point 3 on the cardiac graph). Then the atria fill with blood again due to higher pressure in the vena cava and pulmonary vein and the cycle starts over again.

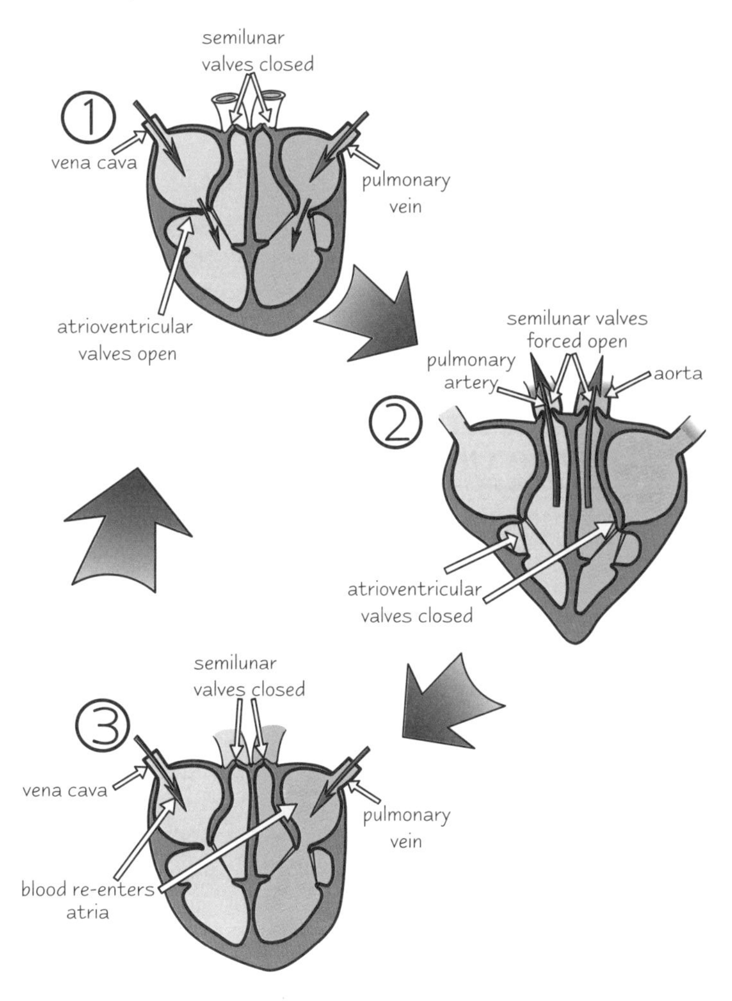

Practice Questions

Q1 Which side of the heart carries oxygenated blood?

Q2 Why is the left ventricle more muscular than the right ventricle?

Q3 What is the purpose of heart valves?

Exam Questions

Q1 Describe the pressure changes which occur in the heart during systole and diastole. [3 marks]

Q2 Explain how valves stop blood going back the wrong way. [6 marks]

Learn these pages off by heart…

Some of this will be familiar to you from GCSEs — so there's no excuse for not learning it really well.
The diagram of the heart can be confusing — it's like looking at a mirror image, so right is left and left is right.
(So in fact, when you look in the mirror you don't see what you actually look like — you see a reverse image — weird.)

Transport Systems

The idea behind this page is simple. You take stuff in through specially adapted exchange organs. You move it around to where it's needed in specially adapted transport systems. Then you move waste to where it can be got rid of.

Multicellular Organisms need Transport Systems

All cells need energy — most cells get energy via **aerobic respiration**. The raw materials for this are **glucose** and **oxygen**, so the body has to make sure it can deliver enough of these to all its cells.
In single-celled creatures, these materials can **diffuse directly** into the cell across the cell surface membrane (see p.22). The diffusion rate is quick because of the small distances the substances have to travel (see p.18).

In **multicellular** animals, diffusion across the outer membrane is too slow for their needs. This is because:

1) some cells are **deep within the body**;
2) they have a **low surface area to volume ratio**;
3) they have a **high metabolic rate**, which means they respire quickly, so they need a **constant supply** of glucose and oxygen.
4) they have a **tough outer surface**.

So multicellular animals need **transport systems** to carry raw materials from specialised **exchange organs** to their body cells. In mammals this is the **circulatory system**, which uses **blood** to carry glucose and oxygen around the body. It also carries **hormones**, **antibodies** (to fight disease) and **waste** like CO_2. The bulk movement of blood is called **mass flow**.

Exchange Organs are Specialised for their Function

For **exchange surfaces** to be efficient, they need to have:

1) thin membranes to provide a **short diffusion path**;
2) a **big surface area** so they can exchange enough gases or nutrients.

In mammals gas exchange happens in the **alveoli** of the lungs.
Nutrients are absorbed in the **villi** of the small intestine. Both alveoli and villi:

1) transfer substances into **blood capillaries**;
2) have blood capillaries very close to the surface;
3) are extremely small and numerous to give a big **surface area to volume ratio**;
4) have thin walls to speed up **diffusion rates**.

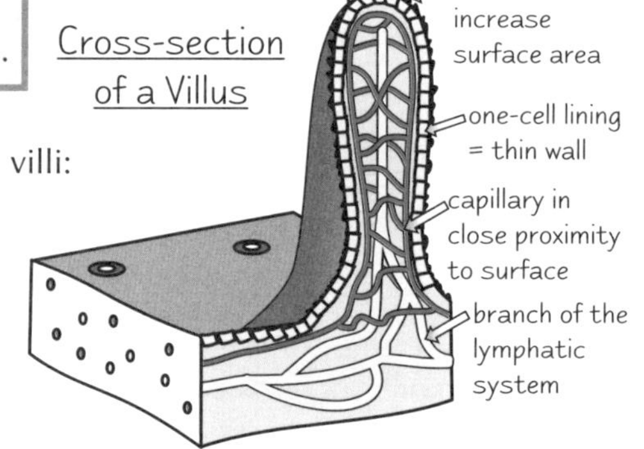

Large Organisms Maintain Concentration Gradients at Exchange Surfaces

If there's a big difference in the concentration of a substance between two areas (a steep **concentration gradient**), the rate of diffusion is faster. Substances diffuse from an area of **higher** concentration to an area of **lower** concentration. One function of the **specialised exchange organs** of larger organisms is to maintain concentration gradients so things diffuse quickly. A good example of this is the **alveoli** in human lungs:

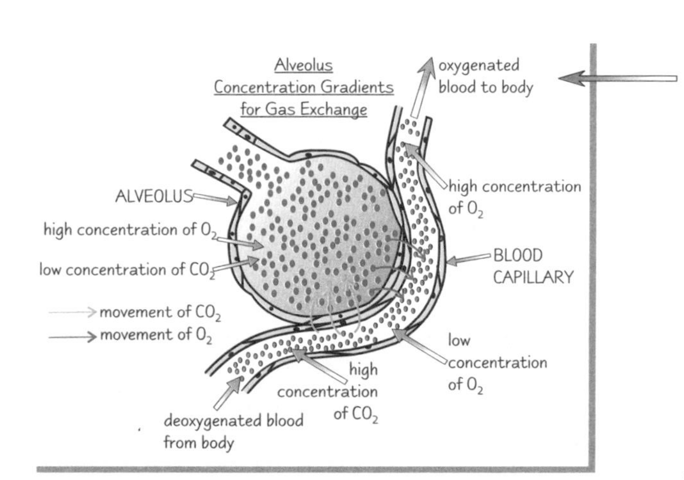

After exchanging gases with cells, **deoxygenated blood** full of CO_2 returns to blood capillaries in the lungs. There's a higher level of $\mathbf{O_2}$ in the **alveoli** compared to the capillaries. This gives a good concentration gradient for **diffusion of oxygen** into the **blood**. The low level of CO_2 in the **alveolar space** helps to remove the CO_2 from the blood. **CO_2 diffuses** down the concentration gradient and **into the lungs**, where it's breathed out.

Other exchange surfaces have similar features. E.g. blood delivers glucose to cells for respiration. When blood returns "empty" to the microvilli in the gut, more glucose diffuses down the concentration gradient into the capillaries. Most larger organisms have special ways of maintaining concentration gradients, like the counter-current exchange system in the gill lamellae of fish (see p.24).

Transport Systems

There are Four Main Types of Blood Vessel

Blood flows around the body in blood vessels.
The four main types of blood vessels are **arteries**, **arterioles**, **capillaries** and **veins**.

1) **Arteries** carry blood under high pressure **from** the heart **to** the rest of the body. They're thick-walled, muscular and have elastic tissue in the walls to cope with the **high pressure** caused by the heartbeat. All arteries carry **oxygenated** blood except the **pulmonary arteries**, which take deoxygenated blood to the lungs.

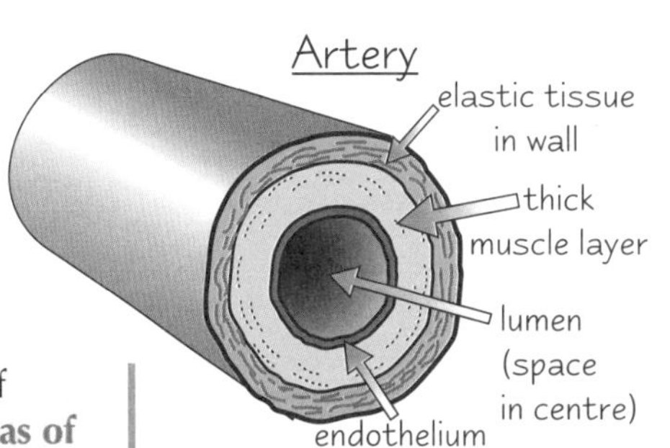

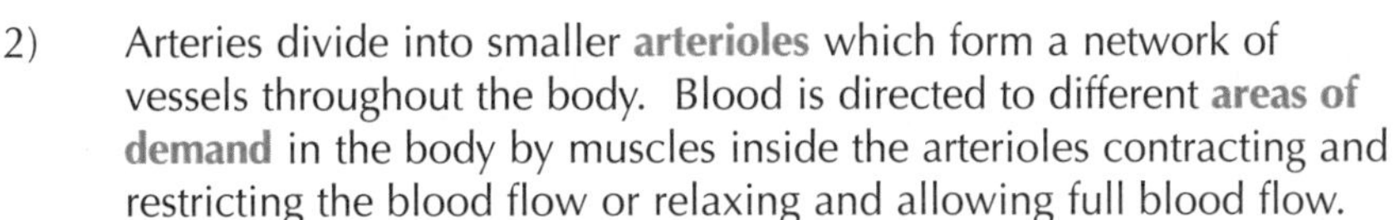

2) Arteries divide into smaller **arterioles** which form a network of vessels throughout the body. Blood is directed to different **areas of demand** in the body by muscles inside the arterioles contracting and restricting the blood flow or relaxing and allowing full blood flow.

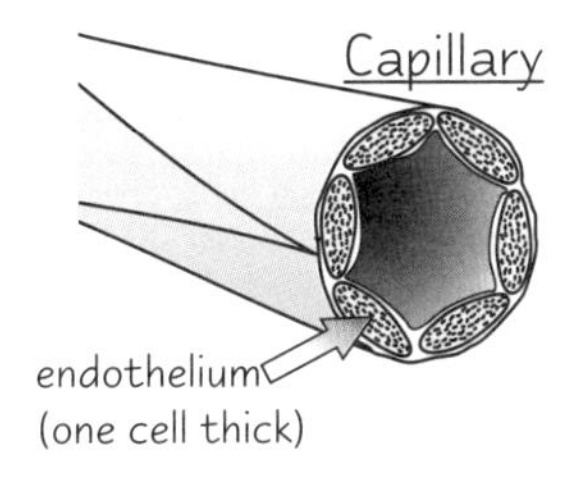

3) Arterioles branch into **capillaries**, which are the **smallest** of the blood vessels. Substances are exchanged between cells and capillaries. Their walls are only **one cell thick** to allow efficient **diffusion** of substances (e.g. glucose and oxygen) to occur near cells. Networks of capillaries in tissue are called **capillary beds**.

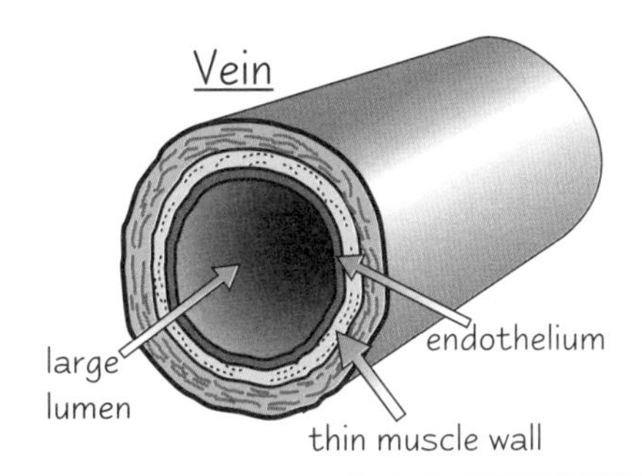

4) **Veins** take low pressure blood back **to the heart**. They're **wider** than equivalent arteries, with very little elastic or muscle tissue. Veins contain **valves** to stop the blood flowing backwards. Blood flow through the veins is helped by contraction of the **body muscles** surrounding them. All veins carry **deoxygenated** blood (because oxygen has been used up by body cells), except for the **pulmonary veins**, which carry oxygenated blood to the heart from the lungs.

Practice Questions

Q1 Why don't single-celled organisms need transport systems?

Q2 How does a concentration gradient influence speed of diffusion?

Q3 What are the 4 main types of blood vessel?

Q4 Do arteries mainly carry oxygenated or deoxygenated blood?

Q5 What is the smallest type of blood vessel?

Exam Questions

Q1 "Structures involved in exchange are specialised for their function." Give two examples of exchange structures in the human body and describe the relationship between their structure and their function. [4 marks]

Q2 Why do multicellular organisms need extra features to aid diffusion of substances into and out of their cells? [6 marks]

If blood can handle transport this efficiently, the trains have no excuse…

Four hours I was waiting at Preston this weekend. Four hours! Anyway, you may have noticed that biologists are obsessed with the relationship between structure and function, so whenever you're learning the structure of something, make sure you know how this relates to its function. Like the veins, arteries, arterioles and capillaries on this page, for example.

Blood, Tissue Fluid and Lymph

For some people the word 'blood' is enough to make them cringe. Strange really, seeing as it's the substance that keeps us all alive. So sorry to all you blood-phobics out there, because these pages are all about the stuff.

Blood Contains Blood Cells, Platelets and Plasma

Blood is a **specialised tissue** that's composed of 45% **blood cells** and **platelets** suspended in a liquid called **plasma** (55%). Plasma is mainly **water**, with various nutrients and gases dissolved in it. The main role of blood is **transporting substances** around the body **dissolved** in the **plasma**. Substances move into and out of the plasma through **blood capillaries** at exchange surfaces.

Substance	Exchange surface where substance enters blood through capillaries	Exchange surface where substance leaves blood through capillaries
nutrients from digestion (glucose, amino acids, fatty acids, mineral ions)	epithelium of the villi in the small intestine	body tissues
hormones	glands	target organs
oxygen	alveoli in the lungs	body tissues
carbon dioxide	body tissues	alveoli in the lungs
urea	liver cells	kidney cells

Also, antibodies are secreted directly into plasma by white blood cells

Blood Cells are Adapted to Specific Jobs

There are three main types of blood cell:

1) **Red blood cells** (**erythrocytes**) are responsible for absorbing **oxygen** and transporting it round the body. They're made in the **bone marrow** and are very small.

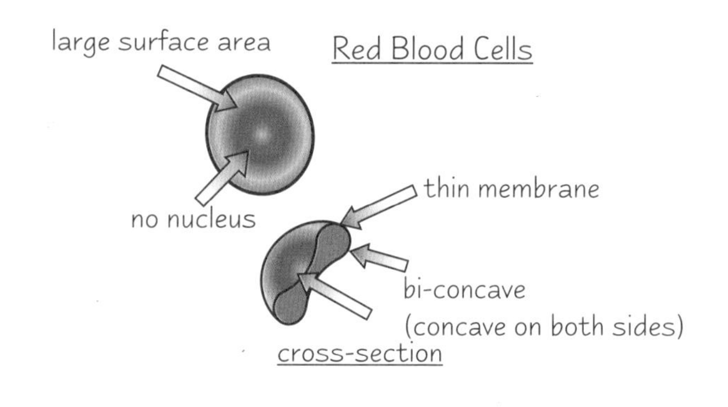

1) They have **no organelles** (including no nucleus) to leave more room for **haemoglobin**, which carries the oxygen.
2) They have a large surface area due to their **bi-concave disc** shape. This allows O_2 to diffuse quickly into and out of the cell.
3) They have an **elastic membrane**, which allows them to change shape to squeeze through the small blood capillaries, then spring back into normal shape when they re-enter veins.

2) **White blood cells** (**leucocytes**) are larger than red blood cells but there are fewer of them in blood. They are responsible for fighting disease.

3) **Platelets** are tiny bits of **cytoplasm** held within a cell membrane. They have no nucleus. They release an enzyme that produces **fibrin fibres** to form a **blood clot** when blood vessels are cut open — this is what forms **scabs**.

Blood, Tissue Fluid and Lymph

Tissue Fluid is Formed from Blood Plasma

Tissue fluid surrounds the cells in tissues — providing them with the conditions they need to function. Tissue fluid is made from substances which leave the plasma from the blood capillaries. Substances move out of blood capillaries due to a **pressure gradient**:

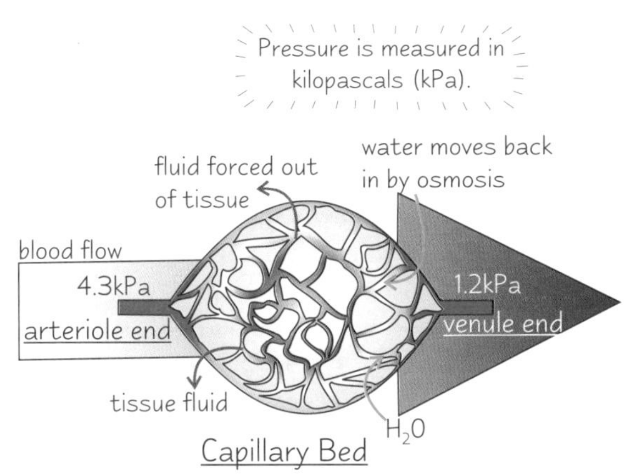

1) At the **arteriole end** of the capillary bed, pressure inside the capillaries is **greater** than pressure in the tissue fluid. This difference in pressure forces fluid to **leave** the **capillaries** and enter tissue space.
2) As fluid leaves, pressure is reduced in the capillaries — so the pressure is much lower at the **venule end** of the capillary bed.
3) Due to the fluid loss, the **water potential** at the **venule end** of the capillaries is **lower** than the water potential in the **tissue fluid** — so some **water re-enters** the capillaries from the tissue fluid at the venule end, by **osmosis**.

Unlike blood, tissue fluid **doesn't** contain **red blood cells** or **big proteins**, because they are **too large** to be pushed out through the capillary walls. It does contain smaller molecules, e.g. oxygen, glucose and mineral ions. Tissue fluid helps cells to get the oxygen and glucose they need, and to get rid of the CO_2 and waste they don't need.

Lymph is Formed from Excess Tissue Fluid

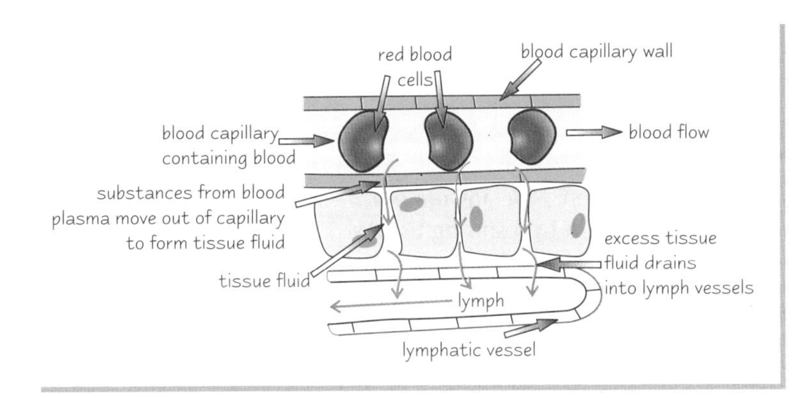

Lymph is a fluid that forms when excess tissue fluid drains into the lymphatic vessels, which lie close to blood capillaries. Lymph takes away waste products from the cells. It then travels through the **lymphatic system** and eventually enters blood plasma.

It's similar to tissue fluid, except it contains more fats, proteins and white blood cells, which it picks up at **lymph nodes** as it travels through the lymphatic system.

Practice Questions

Q1 What does blood contain?

Q2 What do white blood cells do?

Q3 What are all cells surrounded by?

Q4 Why doesn't tissue fluid contain red blood cells?

Q5 What makes lymph different from tissue fluid?

Exam Questions

Q1 How does tissue fluid move into and out of blood capillaries? [5 marks]

Q2 Describe three ways that red blood cells are adapted for their function. [3 marks]

That's the end of this bloody topic — and you can't get me for swearing...

It's time to conquer any fears of blood and start appreciating it for the amazing tissue it is. Learn its functions, and which bits do each function — plasma does transport, red blood cells do oxygen and white blood cells do disease-fighting. If I were a blood cell I'd be a great leucocyte warrior, instilling fear into the nuclei of pathogens everywhere. Aaanyway.

Haemoglobin and Oxygen Transport

Aaagh, complicated topic alert. Don't worry though, because your poor, over-worked brain cells will recover from the brain-strain of these pages thanks to oxyhaemoglobin. So the least you can do is learn how it works.

Oxygen is Carried Round the Body as *Oxyhaemoglobin*

Oxygen is carried round the body by **haemoglobin** (Hb), in red blood cells. When oxygen joins to haemoglobin, it becomes **oxyhaemoglobin**. This is a **reversible reaction** — when oxygen leaves oxyhaemoglobin (**dissociates** from it), it turns back to haemoglobin.

Hb	**+ $4O_2$** $\rightleftharpoons$	**HbO_8**
Haemoglobin	+ oxygen $\rightleftharpoons$	oxyhaemoglobin

1) **Haemoglobin** is a large, **globular protein** molecule made up of four polypeptide chains.
2) Each chain has a **haem group** which contains **iron** and gives haemoglobin its **red** colour.
3) Haemoglobin has a **high affinity for oxygen** — each molecule carries **four oxygen molecules**.

'Affinity' for oxygen means willingness to combine with oxygen.

Partial Pressure Measures *Concentration* of *Gases*

The **partial pressure of oxygen (*p*O_2)** is a measure of **oxygen concentration**.
The **greater** the concentration of dissolved oxygen in cells, the **higher** the partial pressure.
Similarly, the **partial pressure of carbon dioxide (*p*CO_2)** is a measure of the concentration of carbon dioxide in a cell.

Oxygen **loads onto** haemoglobin to form oxyhaemoglobin where there's a **high *p*O_2**.
Oxyhaemoglobin **unloads** its oxygen where there has been a **decrease in *p*O_2**.

1) Oxygen enters blood capillaries at the **alveoli** in the **lungs**. Alveoli cells have a **high *p*O_2** so oxygen **loads onto** haemoglobin to form oxyhaemoglobin.
2) When our **cells respire**, they use up oxygen. This **lowers *p*O_2**, so red blood cells deliver oxyhaemoglobin to respiring tissues, where it unloads its oxygen.
3) The haemoglobin then returns to the lungs to pick up more oxygen.

Dissociation Curves Show How *Affinity for Oxygen* Varies

Dissociation curves show how the willingness of haemoglobin to combine with oxygen varies, depending on partial pressure of oxygen (*p*O_2).

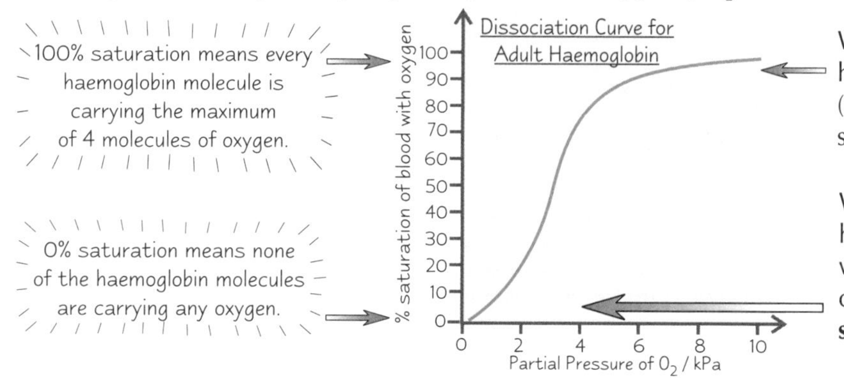

Where ***p*O_2 is high** (e.g. in the lungs), haemoglobin has a **high affinity** for oxygen (i.e. it will **readily combine** with oxygen), so it has a **high saturation** of oxygen.

Where ***p*O_2 is low** (e.g. in respiring tissues), haemoglobin has a **low affinity** for oxygen, which means it **releases oxygen** rather than combines with it. That's why it has a **low saturation** of oxygen.

The graph is **'S-shaped'** because when haemoglobin (Hb) combines with the **first O_2 molecule**, it **alters the shape** of the Hb molecule in a way that makes it **easier** for other molecules to join too. But as the haemoglobin starts to become fully saturated, it becomes harder for more oxygen to join. As a result, the curve has a **steep** bit in the middle where it's really easy for oxygen molecules to join, and **shallow** bits at each end where it's harder for oxygen molecules to join.
When the curve is steep, a small change in pO_2 causes a big change in the amount of oxygen carried by the haemoglobin.

Haemoglobin and Oxygen Transport

Carbon Dioxide Levels Affect Oxygen Unloading

To complicate matters, haemoglobin gives up its oxygen **more readily** at **higher partial pressures of carbon dioxide** (pCO_2). It's a cunning way of getting more oxygen to cells during activity. When cells respire they produce carbon dioxide, which raises pCO_2, increasing the rate of oxygen unloading. The reason for this is linked to how CO_2 affects blood pH.

1) CO_2 from respiring tissues diffuses into red blood cells and is converted to **carbonic acid**.
2) The carbonic acid **dissociates** to give **hydrogen ions** and **hydrogencarbonate ions**.
3) If left alone, the hydrogen ions would increase the cell's acidity. To prevent this, oxyhaemoglobin **unloads** its oxygen so that haemoglobin can take up the hydrogen ions.
4) The **hydrogencarbonate ions** diffuse out of the red blood cells and are **transported in the plasma**.
5) When the blood reaches the **lungs** the low concentration of CO_2 causes the hydrogencarbonate and hydrogen ions to **recombine into CO_2**.
6) The CO_2 then diffuses into the **alveoli** and is breathed out.

When carbon dioxide levels increase, the dissociation curve 'shifts' down, showing that more oxygen is released from the blood (because the lower the saturation of O_2 in blood, the more O_2 is being released). This is called the Bohr effect.

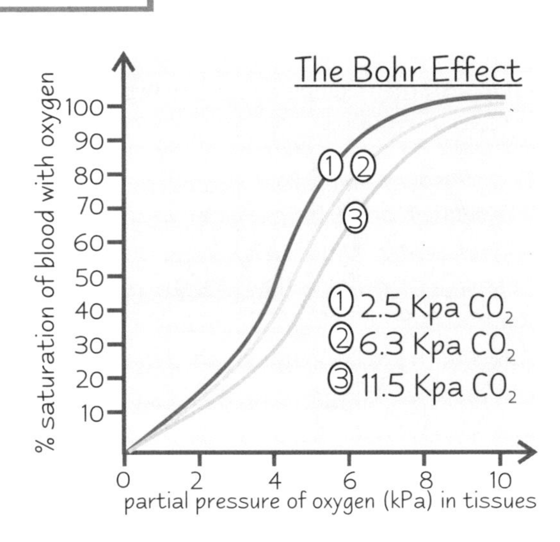

Practice Questions

Q1 How is HbO_8 formed?

Q2 What is pO_2?

Q3 What is carbon dioxide converted to in red blood cells?

Q4 What does affinity mean?

Exam Question

Q1 Why is the dissociation curve for haemoglobin "S-shaped"? [3 marks]

The Bore effect — it's happening right now…

Dissociation graphs can be a bit confusing — but basically, when tissues contain lots of oxygen (i.e. pO_2 is high), haemoglobin readily combines with the oxygen, so blood has a high saturation of oxygen (and vice versa when pO_2 is low). Simple. Also, make sure you get the lingo right, like 'partial pressure' and 'affinity' — hey, I'm hip, I'm groovy.

Control of Heartbeat

You don't have to think consciously about making your heart beat — your body does it for you. So you couldn't stop it beating even if for some strange reason you wanted to. Which is nice to know.

Cardiac Muscle Controls the **Regular Beating** of the Heart

Cardiac muscle is '**myogenic**' — this means that, rather than receiving signals from **nerves**, it contracts and relaxes on its own. This pattern of contractions controls the **regular heartbeat**.

1) The process starts in the **sino-atrial node (SAN)** in the wall of the **right atrium**.
2) The SAN is like a pacemaker — it sets the rhythm of the heartbeat by sending out regular **electrical impulses** to the atrial walls.
3) This causes the right and left atria to contract **at the same time**.
4) A band of non-conducting **collagen tissue** prevents the electrical impulses from passing directly from the atria to the ventricles.
5) Instead, the **atrio-ventricular node (AVN)** picks up the impulses from the SAN. There is a **slight delay** before it reacts, so that the ventricles contract **after** the atria.
6) The AVN generates its own **electrical impulse**. This travels through a group of fibres called the **bundle of His** and then into the finer fibrous tissue in the right and left ventricle walls called **Purkyne tissue**.
7) The impulses mean both ventricles **contract simultaneously**, from the bottom up.

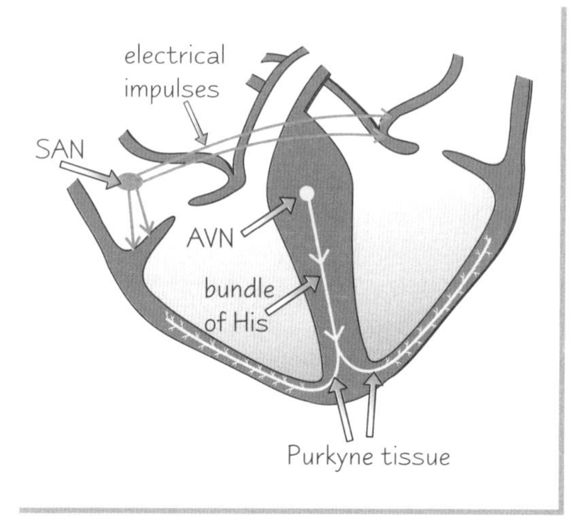

The **Brain** Controls **Changes** in the **Heart Rate**

Nerve impulses from the brain **modify the heart rate** — making the heart beat faster or slower.

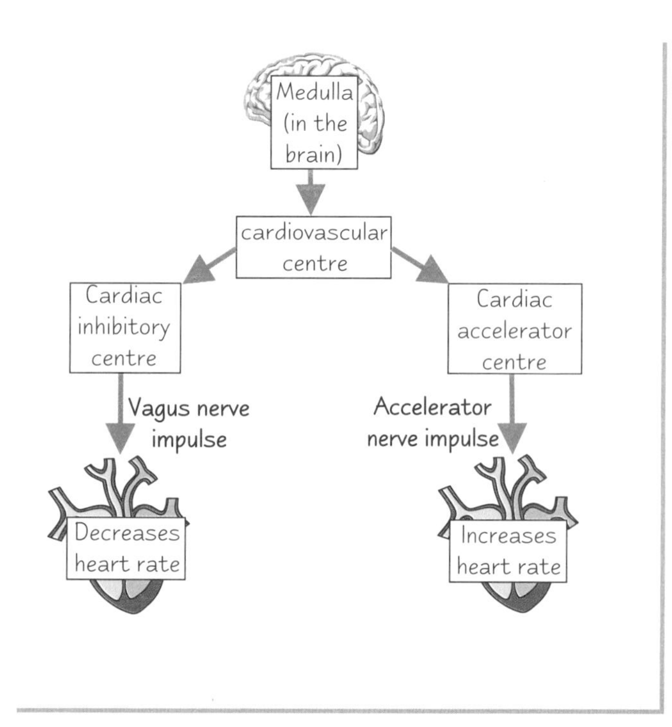

1) **Heart rate** is controlled in the **cardiovascular centre** in the **medulla oblongata** area of the **brain**.
2) **Nerve impulses** from the **cardiovascular centre** reach the SAN.
3) An **accelerator nerve impulse** from the cardiac accelerator centre stimulates the SAN to **increase the heartbeat**.
4) A **vagus nerve impulse** from the cardiac inhibitory centre stimulates the SAN to **slow the heart rate down**.

Doug's medulla oblongata was having trouble steadying his heart rate.

Control of Heartbeat

Receptors Inform the Brain About **Changing Blood Pressure** and **Blood pH**

Pressure receptors inform the brain about changes in **blood pressure** so it can alter heart rate accordingly:

1) It's important that **blood pressure** is maintained at a constant level. **Pressure receptors** are found in the **aorta wall** and in **carotid arteries**. They detect changes in arterial blood pressure and inform the brain.
2) If the pressure is **too high**, they send impulses to the cardiovascular centre. In turn, this sends impulses to the SAN via the **parasympathetic nervous system**, to **slow down heart rate**.
3) If **pressure is too low**, pressure receptors send impulses to the cardiovascular centre, which sends its own impulses to, yep you guessed it, the SAN via the **sympathetic nervous system**, to **speed up heart rate**.

Chemoreceptors inform the brain about changes in **blood pH** so it can alter heart rate accordingly:

1) An **increase in respiration**, for example due to physical activity, increases the **level of CO_2** in the blood. Increased CO_2 levels **lower the pH of blood**, which is detected by **chemoreceptors**, found in **carotid sinuses** in the **carotid artery walls**.
2) In the same way as pressure receptors, these send impulses to the brain, which sends its own impulses to the SAN to **increase the heart rate**.

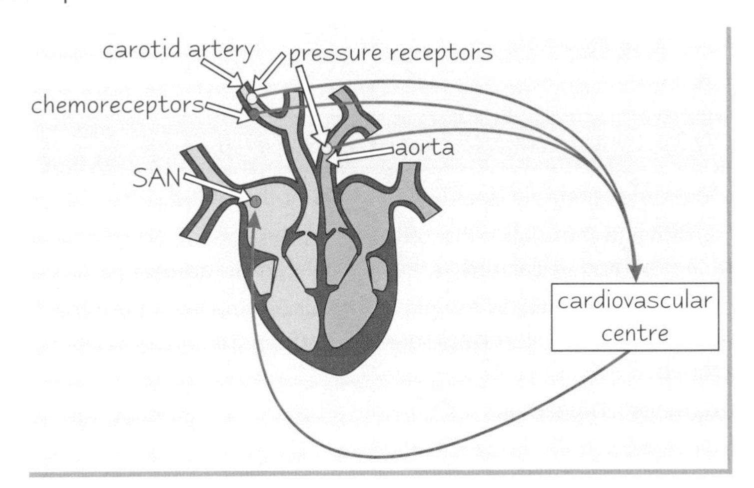

Practice Questions

Q1 What does myogenic mean?

Q2 What is the difference between the SAN and the AVN?

Q3 What two types of tissue do electrical impulses from the AVN travel through?

Q4 Which part of the brain modifies heart rate?

Q5 How do pressure receptors control blood pressure?

Exam Questions

Q1 Describe what would happen to the heart rate if the heart suffered a sudden loss of blood pressure. [4 marks]

Q2 Describe how the heartbeat is stimulated. [6 marks]

What happened to the Bundle of Hers? — she lost her nerve & had to go…

Did you know that your heart beat can continue beating for a while after you're 'clinically dead' — spooky stuff. The reason is that it doesn't need any input from the brain and nervous system to operate. But, although the brain doesn't control heartbeat it does control the ***rate*** *of the heartbeat. That's where receptors and the cardiovascular centre come in.*

Ventilation and Types of Respiration

A person's breathing rate changes when they exercise. Just like heart rate it's all controlled by some receptors and the medulla in the brain. Vigorous exercise also means that cells start anaerobic respiration.

*The **Medulla Oblongata** in the Brain Controls **Rate of Breathing***

Your **breathing rate** changes according to how much **physical activity** you're doing. When you're exercising you use more **energy**. Your body needs to do more **aerobic respiration** to release this energy, so it needs more **oxygen**.

The **ventilation cycle** is the cycle of breathing in and out. It involves **inspiratory** and **expiratory** centres in the **medulla oblongata** (an area of the brain) and **stretch receptors** in the lungs.

1) The medulla's **inspiratory centre** sends nerve impulses to the **intercostal** (rib) and **diaphragm** muscles to make them **contract**. It also sends nerve impulses to the medulla to **inhibit** the **expiratory centre**.
2) Air enters the lungs due to the **pressure difference** between the lungs and the air outside.
3) The **lungs inflate**. This stimulates **stretch receptors**, which send nerve impulses back to the **medulla** to **inhibit** the **inspiratory centre**.
4) Now the expiratory centre (no longer inhibited) sends nerve impulses to the muscles to relax and the **lungs deflate**, expelling air. This causes the **stretch receptors** to become **inactive**, so the inspiratory centre is no longer inhibited and the cycle starts again.
5) This ventilation cycle happens **automatically** without you having to think about it.

*The Medulla Responds to **Chemical Changes** in the Blood*

1) During exercise, CO_2 levels rise and this **decreases** the **pH** of the blood.
2) **Chemoreceptors** are sensitive to these chemical changes in the blood. They're found in the **medulla oblongata**, in **aortic bodies** (in the aorta), and in **carotid bodies** (in the carotid arteries carrying blood to the head).
3) If the chemoreceptors **detect** a **decrease** in the **pH** of the blood, they send a **signal** to the **medulla** to send more frequent nerve impulses to the intercostal muscles and diaphragm. This **increases** the **rate** of **breathing** and the **depth** of breathing.
4) This allows **gaseous exchange** to **speed up**. CO_2 levels drop and the demand for extra O_2 by the muscles is met.

*Energy for Exercise comes from **Respiration***

When you exercise you **respire** more to get enough energy. Respiration is the release of energy from food. **Glucose** is the main substance used for respiration, but if it's not available the body can use **glycogen** (stored energy) or **triglycerides**. The energy released by respiration is then stored as **ATP**, a **short-term** energy store.

There are two kinds of respiration — aerobic and anaerobic. Aerobic respiration uses oxygen. Anaerobic respiration takes over when oxygen runs out — for example when you're exercising hard.

Aerobic Respiration —

1) Uses oxygen and produces waste CO_2 that's excreted through the lungs.
2) Releases more energy from each glucose molecule than anaerobic respiration.

$$C_6H_{12}O_6 + 6O_2 \longrightarrow 6CO_2 + 6H_2O + 38\ ATP$$

glucose + oxygen ⟶ carbon dioxide + water + ATP (contains energy)

Anaerobic Respiration —

1) In humans it's called **lactate fermentation**.
2) It's less efficient at releasing energy.
3) It doesn't need oxygen to release energy.
4) It produces **lactic acid**, which builds up in the blood. This lowers the pH in muscle cells that are respiring anaerobically, which causes the pain known as **muscle fatigue**.

$$C_6H_{12}O_6 \longrightarrow 2CH_3CHOHCOOH + 2ATP$$

glucose ⟶ lactic acid + ATP (contains energy)

Ventilation and Types of Respiration

During Exercise an **Oxygen Debt** Can Build Up

During **vigorous exercise** the body demands **more oxygen** than is available:

1) In skeletal muscles, **aerobic** respiration changes to **anaerobic** respiration.
2) **Lactic acid** is produced, which builds up in the blood. Lactic acid is **toxic** and the body can only deal with **small amounts** — it leads to muscular pain and tiredness. If there is too much lactic acid in the blood, exercise has to stop.
3) The lactic acid is carried in the blood to the **liver**. In the liver, lactic acid is converted to the chemical **pyruvate** and then to **glucose**.
4) The liver needs **oxygen** to get rid of the lactic acid. When exercise stops, more oxygen can be used for this job.
5) Once lactic acid has been converted back to glucose, it can either be taken back to **respiring cells** to be used as energy or be stored as **glycogen**.
6) The **oxygen** needed to **convert lactic acid** to pyruvate and then to glucose is called the **oxygen debt**. This is why you keep **panting** after hard **exercise** — you're still **repaying** the oxygen debt.

Aerobic exercise** is **Way** more **Efficient

Aerobic exercise is exercise using oxygen. If you pace yourself well, you can have a good work-out without going into anaerobic respiration. Athletes train regularly to train their muscles to work harder whilst staying in aerobic respiration (which provides more energy). Aerobic exercise improves **ventilation** and makes the **circulatory system** more efficient.

Practice Questions

Q1 Explain the role of stretch receptors in the ventilation cycle.

Q2 What structures detect changes in the pH of blood?

Q3 Where is the rate of breathing controlled?

Q4 What is the main substance used for respiration in animals?

Q5 Write out both the word and formula equations for aerobic and anaerobic respiration.

Q6 Where is lactic acid converted to glucose?

Q7 What does the term "oxygen debt" mean?

Exam Questions

Q1 Explain how chemoreceptors and the medulla oblongata are able to bring about an increase in breathing rate in response to exercise. [4 marks]

Q2 Describe what happens to the lactic acid that is produced by muscles during anaerobic respiration. [3 marks]

If you don't learn all this stuff now you'll end up with revision debt...

Fair enough there's quite a lot of stuff rammed on these pages but if you think about it it's all quite easy. There's nothing really complicated, it's just a matter of learning a few equations and lots of facts. So shut the book and keep writing till the fat cow sings and the ladies comes home...or at least till you know it off by heart.

Roots, Xylem and Phloem

Water enters a plant through its roots and eventually, if it's not used, exits via the leaves. Plants have two kinds of vessel — xylem and phloem. They're responsible for moving water, ions and nutrients around the plant.

Root Cells** are **Specially Adapted** to **Take Up Water

Water has to get from the **soil**, across the **root** and into the **xylem**, which takes it up the plant. The bit of the root that absorbs water is covered in **root hairs**. This increases its surface area and speeds up water uptake. Once it's absorbed, the water has to get through two root tissues, the **cortex** and the **endodermis**, to reach the xylem.

*There are **Three Routes** Water can Take through the **Root***

Water can then travel through the roots into the xylem by three different paths:

1) The **apoplast pathway** — goes through the **non-living** parts of the root — the **cell walls**. The walls are very absorbent and water can simply diffuse through them, as well as passing through the spaces between them.
2) The **symplast pathway** — goes through the **living** cytoplasm of the cells. The **cytoplasm** of neighbouring cells connects through **plasmodesmata** (small gaps in the cell walls).
3) The **vacuolar pathway** — water travels from the vacuole of one cell into the vacuole of the next by **osmosis**.

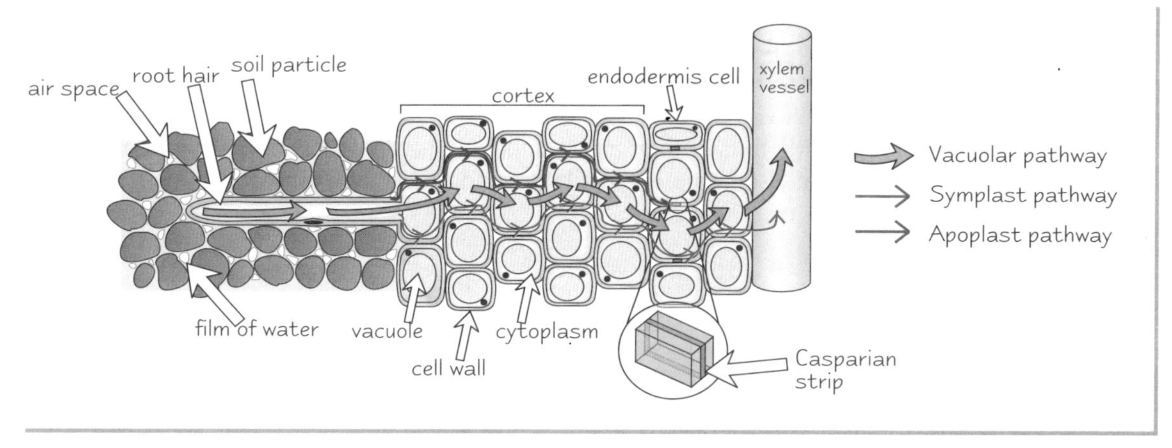

All three pathways are used, but the main one is the **apoplast pathway** because it provides the **least resistance**. When the water gets to the **endodermis** cells, though, the apoplast pathway is blocked by a **waxy strip** in the cell walls, called the **Casparian strip**, which the water can't penetrate. Now the water has to take one of the other pathways. This is useful, because it means the water has to go through a **cell membrane**. Cell membranes are able to control whether or not substances in the water get through (see p.16-17). Once past this barrier, the water moves into the **xylem**.

Xylem** and **Phloem** have **Different Structures** in **Different Parts of the Plant

Xylem and **phloem** are the two **transport** systems within plants:

- **Xylem** carries **water** and **mineral ions**. It's also used for **support**.
- **Phloem** carries **dissolved food**.

They are **distributed** differently in different parts of the plant:

1) In a **root**, which has to resist crushing as it pushes though the soil, the xylem is in the **centre**.
2) In **stems**, which need to resist bending, the xylem is **near the outside** to provide a sort of 'scaffolding'.
3) In a **leaf**, xylem and phloem make up a **network of veins**, which is needed for support because leaves are thin.

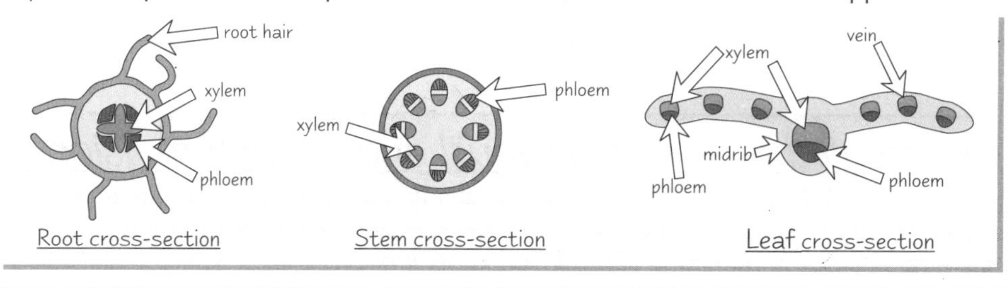

Roots, Xylem and Phloem

Xylem Carries Water and Provides Support

There are two kinds of xylem structure that you need to know about:

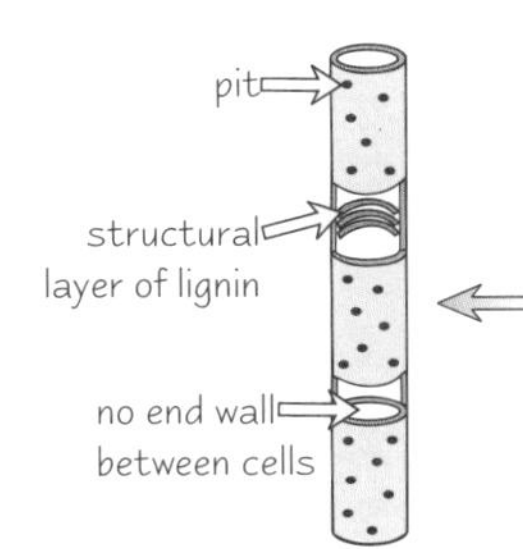

Vessels — These are very long, **tube-like** structures formed from cells (**vessel elements**) joined end to end. There are **no end walls** on these cells, making an **uninterrupted tube** that allows water to pass through easily. The vessels are **dead**, containing **no cytoplasm**. Their walls are thickened with a woody substance called **lignin**, which helps with **support**. The amount of lignin increases as the cell gets older. Substances get into and out of the vessels through small **pits** in the walls, where there is **no lignin**.

Tracheids — Like vessels, these are long, **lignified** cells with no cytoplasm. They don't form continuous tubes like vessels do — instead they **overlap** so that water can pass through. Where cells overlap the wall contains extra pits, so substances can pass easily between cells. These cells are a **more primitive** form of xylem — plants usually have more vessels than tracheids.

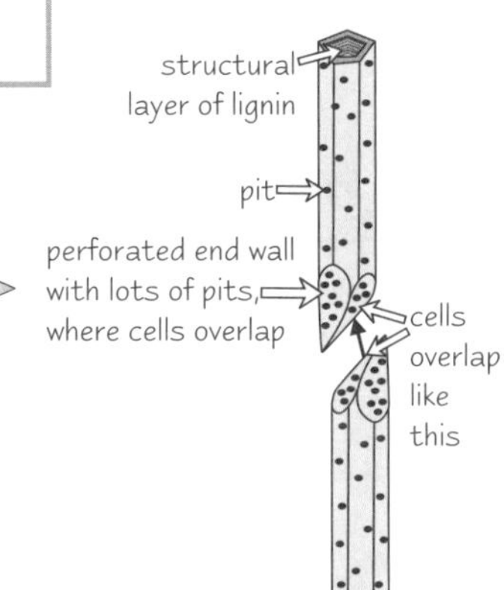

There are Two Main Cell Types in Phloem

Phloem tissue transports **solutes** (dissolved substances — mainly sucrose) round plants. Like xylem, **phloem** is formed from cells arranged in **tubes**, and these cells are modified for transport. But, unlike xylem, it's purely a **transport tissue** — it isn't used for support as well.

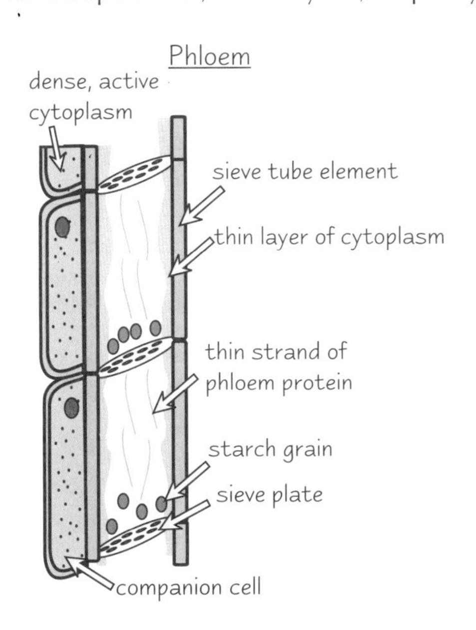

1) **Sieve tube elements** — These are **living cells** that **transport solutes** through the plant. They are joined end-to-end to form **sieve tubes**. The 'sieve' parts are the end walls, which have lots of holes in them. Unusually for living cells, sieve tube elements have **no nucleus** and only a **very thin layer of cytoplasm** without many organelles. The cytoplasm of adjacent cells is connected through the holes in the sieve plates.

2) **Companion cells** — The lack of a nucleus and other organelles in sieve tube elements means that they would have difficulty surviving on their own. So there is a companion cell for every sieve tube element. The companion cell has a very **dense and active cytoplasm**, and it seems to carry out the living functions for both itself and its sieve cell. Both are formed from a single cell during the development of the phloem.

Phloem tissue also contains some other types of cell, but **sieve tube elements** and **companion cells** are the most important cell types for transport.

There are some practice questions about xylem and phloem on p. 65

Don't worry.. just relax and go with the phloem

It's vital your mind doesn't wander on this page, because the structure and functions of some of these cell types are quite similar. It can be easy to get mixed up if you haven't learnt it properly, so take the time now to sort out which cell type does what. You also need to sort out the three ways through the root so you don't end up getting confused in the exam.

Transpiration

Plants can't sing, juggle or tap-dance (as you will hopefully be aware). But they can exchange gases and move water upwards against the force of gravity— how exciting. What makes it all the more thrilling is that they lose water vapour as they do it. Gripping stuff.

Transpiration is Loss of Water from a Plant's Surface

Transpiration is a side effect of photosynthesis. **Photosynthesis** uses up the carbon dioxide and **produces water** and oxygen. Water **evaporates** from the moist cell walls and accumulates in the spaces between cells in the leaf. Then it diffuses out of the **stomata** when they open. This happens because there is a **diffusion gradient** — there's more water inside the leaf than in the air outside.

Four Main Factors Affect Transpiration Rate

The factors below affect transpiration rate. Temperature, humidity and wind (air movement) alter the **diffusion gradient**, but **light** is a bit different:

1) **Light** — Transpiration happens mainly when the stomata are open. In the dark the stomata usually close, so there's little transpiration.
2) **Temperature** — Diffusion involves the movement of molecules. Increasing the temperature speeds this movement up. So as temperature rises, so does transpiration rate.
3) **Humidity** — If the air around the plant is humid, the **diffusion gradient** between the leaf and the air is reduced. This slows transpiration down.
4) **Wind** — Lots of air movement blows away water molecules from around the stomata. This **increases** the diffusion gradient, which increases the rate of transpiration.

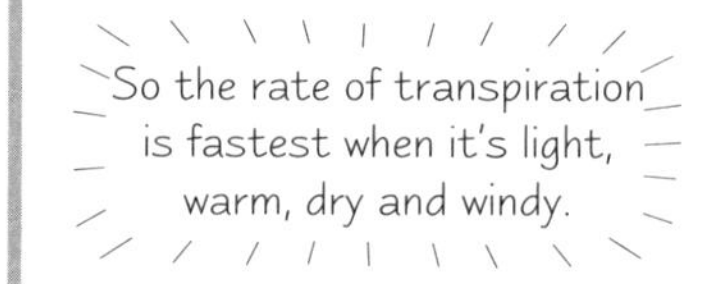

Water Moves Up a Plant Against the Force of Gravity

The **cohesion-tension theory** explains how water moves up plants from roots to leaves, against the force of gravity.

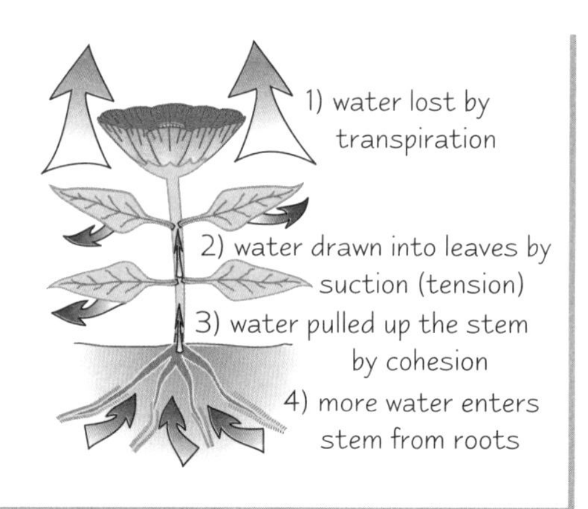

1) Water evaporates from the leaves at the 'top' of the xylem (transpiration).
2) This creates a **suction** ('tension'), which pulls more water into the leaf.
3) Water molecules **stick together** ('cohesion' — see p.10), so when some are pulled into the leaf others follow.
4) This means the whole **column** of water in the xylem, from the leaves down to the roots, moves upwards.

It's like what happens when you suck at the top of a drinking straw and the liquid moves up it.

Root pressure also helps move the water upwards. Water is transported into the xylem in the roots, which creates a pressure and tends to shove water already in the xylem further upwards.
This pressure is weak, and couldn't move water to the top of bigger plants by itself. It helps though, especially in young, small plants where the leaves are still developing.

Transpiration

Plants Need to Absorb **Mineral Ions** as well as Water

Plants absorb **dissolved mineral ions** from the soil. This can happen by **diffusion**, but diffusion **isn't** a selective process — and plants often **only** need **certain ions** and not others. So instead, ions needed by a plant are absorbed by **active transport** (see p.21). This means a plant can absorb more of a certain ion, even if it already has a **high concentration** of them **inside** its cells — so it doesn't have to rely on there being a **diffusion gradient** into the plant. It also means that the plant can **pump out** any ions it **doesn't need**.

Mineral ions travel up the plant with the water, in the **xylem**. Scientists know this because of experiments using **radioactive tracers**. This uses radioactive forms of ions, so that the radioactivity can be detected and the scientist will know where the ion has gone.

Xerophytes are Plants that Live in **Dry Climates**

Xerophytes have adaptations which prevent them losing too much water.
Examples of xerophytic adaptations include:

1) Stomata are sunk in **pits**, where water vapour is sheltered from wind.
2) Leaves are **curled** with the stomata inside, again protecting them from wind.
3) Layer of '**hairs**' on the epidermis traps moist air round the stomata, to reduce the diffusion gradient.
4) **Reduced number of stomata**, so there are fewer places where water can escape.
5) Thick, waxy, water-resistant **cuticle** on the epidermis to reduce water loss.
6) **Swollen stem** to store water.
7) **Roots** spread over a **wide area** just below the soil surface, to make the most of any rain.
8) Whole leaf reduced to a **spike** (e.g. cactus). This reduces the surface area for water loss.
9) **Take in CO_2 at night** so they can close their stomata in the day to reduce water loss, and use stored CO_2 for photosynthesis instead.

Practice Questions (Pages 62-65)

These questions cover the last four pages — don't skip 'em cos you'll miss out on way too much.

Q1 Describe the three routes that water can take through the root.

Q2 What are the functions of xylem tissue?

Q3 How are companion cells and sieve tube elements adapted for their function?

Q4 Define the term "transpiration".

Q5 Write out the stages of the cohesion-tension theory.

Q6 Describe 5 ways that xerophytic plants are adapted to dry conditions

Exam Questions (Pages 62- 65)

Q1 Explain the role of each of the following in the transport of water in the root.
a) cell walls b) the endodermis c) plasmodesmata [6 marks]

Q2 Describe the distribution of the xylem and phloem in stems, roots and leaves.
Explain how this distribution is linked to a function of the xylem. [10 marks]

Q3 Describe factors that can alter the rate of transpiration in a plant, and explain the effect of each. [8 marks]

I don't get it — what's stomata with me?

Actually, that was just a bad joke. Most of it makes sense. The only tricky bit is this diffusion gradient business. All it means is that there's a difference between how many water molecules there are inside the leaf, compared to how many outside. If there aren't many outside but loads inside, they start rushing out to make up for it. The little devils.

Translocation

Translocation *is the movement of* ***organic solutes*** *through a plant. It happens in the* ***phloem****.*
Annoyingly, translocation sounds a lot like transpiration. Or is that just me? Don't confuse them anyway.

The Main Things **Translocated** are **Amino Acids** and **Sugars**

1) Sugars (mostly sucrose) are transported from the leaves (where they're made during photosynthesis) to actively growing regions, or to storage sites.
2) Amino acids are made in the root tips (where nitrogen is absorbed), and are carried to growing areas in the plant to make proteins.

The way that they're transported in the phloem isn't known exactly, but it is known that it's an active process, needing energy from respiration.

Translocation is the movement of dissolved organic substances (mainly sucrose) to where they're needed in the plant. Experiments show that it happens in the phloem (see section below).

Translocation moves substances from 'sources' to 'sinks'. The source of a substance is where it's made (so it's in high concentration there). The sink is the area where it's used up (so it's in low concentration there). For example, the source for sugars is the leaves, and the sinks are the other parts of the plant, especially the food storage organs and growing points in roots, stems and leaves.

Enzymes maintain a **concentration gradient** from the phloem to the sink by **modifying** the organic substances at the sink. For example, in **potatoes**, sucrose is converted to **starch** in the sink areas, so there's always a lower concentration of sucrose at the sink than inside the phloem. This makes sure a **constant supply** of new sucrose reaches the sink from the phloem.

The **Mass Flow Hypothesis** Best Explains **Phloem Transport**

It's still not certain exactly how the solutes are transported from source to sink by translocation. The best supported theory is the **mass flow hypothesis**.

The mass flow hypothesis:

1) Dissolved sugars from photosynthesis (e.g. sucrose) are **actively transported** into the **sieve tubes** of the phloem at the **leaves** (the **source**). This **lowers the water potential** inside the sieve tubes, so water enters the tubes by **osmosis**. This creates a **high pressure** inside the sieve tubes at the source end of the phloem.
2) At the **sink** end, **sugars leave** the phloem to be used up, which **increases the water potential** inside the sieve tubes, so water also leaves the tubes by **osmosis**. This **lowers the pressure** inside the sieve tubes.
3) The result is a **pressure gradient** from source end to sink end, which pushes sugars along the sieve tubes to where they're needed.

Experiments Show **Where** Substances Move to by **Translocation**

Experiments done with **radioactive tracer substances** show how organic **solutes** (**dissolved** organic substances) **move around** in plants. **Radioactive sucrose** is put on a leaf and moves through the plant. This is detected by putting the plant in contact with **photographic film**. Wherever the radioactive sucrose goes, it can be 'seen', because it **fogs** the film. The experiment below shows that, unlike water and mineral ions, which **only move upwards** in the xylem, sucrose moves both **up and down** from the leaf it was added to. This movement can be stopped by **removing a ring of phloem** from the stem. The xylem is still there, so the experiment shows that translocation occurs in the phloem and not in the xylem.

Experiment Showing:
A) that sucrose moves up and down the plant,
B) that translocation occurs in the phloem, not the xylem.

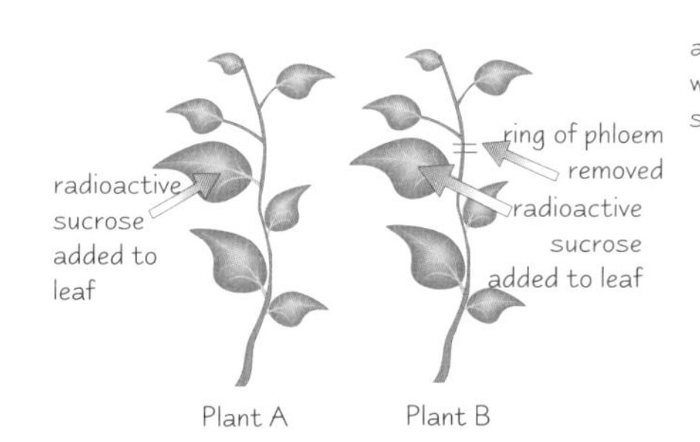

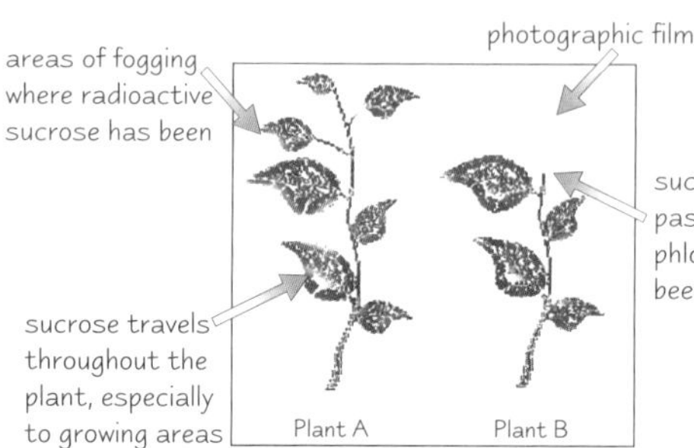

Translocation

Mass Flow Hypothesis Can be Demonstrated in an Experiment

The idea behind the hypothesis can be shown in the experiment below.
In this model, **A** and **B** are two containers, each lined with a **selectively permeable membrane**.

1) **A** represents the **source** end and contains a **concentrated sugar solution**.
2) **B** represents the **sink** end, where **use** of the sugar **lowers** its concentration.
3) Water enters **A** by **osmosis**, causing the sugar solution to flow along the **top tube** — which represents the **phloem**.
4) **Hydrostatic pressure** increases in **B**, forcing water out and back through the **connecting tube** — which represents the **xylem**, because it just transports water.

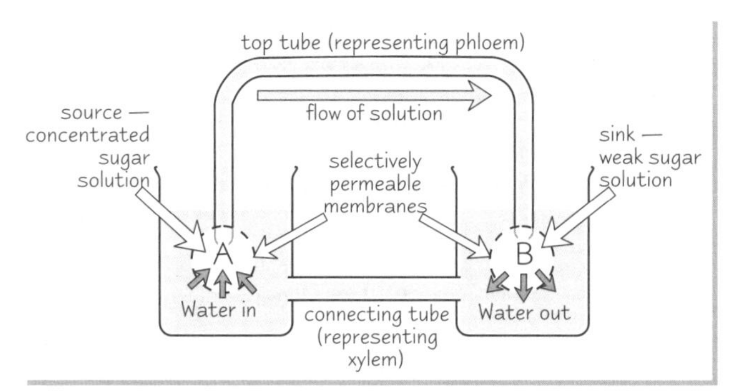

In the model, the **flow stops** when the sugar concentrations in the two containers **equal out**.
But in a plant this **wouldn't happen**, as sugar would constantly be produced by the source and used up by the sink.

Supporting evidence for this theory includes —

1) There is a suitable **water potential gradient** between the leaves and the other parts of the plant.
2) If the phloem is cut sap oozes out, showing that a **pressure gradient** does exist.

Objections to the theory are —

1) Sugar travels to many different sinks, not just to the one with the **highest water potential**, as the model would suggest.
2) The sieve plates would create a barrier to mass flow. A lot of pressure would be needed for the sugar solution to get through.
3) Mass flow doesn't require living cells, yet the **phloem cells** are alive and very active.

This book ain't over 'til the fat pig sings.

Practice Questions

Q1 Explain the terms "source" and "sink" in connection with translocation.

Q2 How can you detect the location of radioactive tracers in a plant?

Q3 State two pieces of evidence that support the mass flow hypothesis for translocation.

Exam Questions

Q1 Radioactive glucose is placed on the leaf of a plant, which is then left for 24 hrs. The radioactivity is then detected by placing the plant in contact with photographic film. In which areas of the plant would you expect to see most radioactivity? Give reasons for your answer. [2 marks]

Q2 The mass flow hypothesis depends on there being a difference in pressure in the phloem sieve tubes between the source and the sink. Explain how sugars cause the pressure to increase at the source end, according to the mass flow hypothesis. [4 marks]

Who cares whether the hypothesis is right — it's the last page...

Blimey it's the end of the book. A sad, sad moment for biology students everywhere. But hey, at least the pig's happy. He's still alive at the end of it all — he's survived cloning, coronary heart disease and xerophytic conditions, amongst everything else. So I think we should all join in for a grand finale sing-song. OK, maybe not — I think I need to go now.

Answers

Section 1 — Core Principles

Page 3 — Carbohydrates

1 *Maximum of 7 marks available.*
Glycosidic bonds are formed by condensation reactions ***[1 mark]*** *and broken by hydrolysis reactions* ***[1 mark]****.*
When a glycosidic bond is formed in a condensation reaction, a hydrogen ***[1 mark]*** *from one monosaccharide combines with a hydroxyl / OH group* ***[1 mark]*** *from the other to form a molecule of water* ***[1 mark]****.*
A hydrolysis reaction is the reverse of this ***[1 mark]****, with a molecule of water being used up to split the monosaccharide molecules apart* ***[1 mark]****.*
The last 5 marks for this question could be obtained by a diagram showing the reaction, using structural formulae.

2 *Maximum of 8 marks available.*
Glycogen's chain is compact and very branched ***[1 mark]*** *whereas cellulose's chain is long, straight and unbranched* ***[1 mark]*** *and these chains are bonded together to form strong fibres* ***[1 mark]****.*
Glycogen's structure makes it a good food store in animals ***[1 mark]****.*
The branches allow enzymes to access the glycosidic bonds to break the food store down quickly ***[1 mark]****.*
Cellulose's structure makes it a good supporting structure in cell walls ***[1 mark]****. The fibres provide strength* ***[1 mark]****.*
The function is helped by the fact that the cell doesn't have any enzymes that can break down the bonds in cellulose ***[1 mark]****.*
In questions worth lots of marks make sure you include enough details. This question is worth 8 marks so you should include at least 8 relevant points to score full marks. Also, the question asks you to compare and contrast, so make sure you don't just describe glycogen and cellulose totally separately from each other. You need to highlight how they differ from each other, and what this means for their functions.

Page 5 — Proteins

1 *Maximum of 7 marks available.*
Proteins are made from amino acids ***[1 mark]****.*
The amino acids are joined together in a long (polypeptide) chain ***[1 mark]****.*
The sequence of amino acids is the protein's primary structure ***[1 mark]****.*
The amino acid chain / polypeptide coils in a certain way ***[1 mark]****.*
The way it's coiled is the protein's secondary structure ***[1 mark]****.*
The coiled chain is itself folded into a specific shape ***[1 mark]****.*
This is the protein's tertiary structure ***[1 mark]****.*

2 *Maximum of 6 marks available.*
Collagen is a fibrous protein ***[1 mark]****.*
For this mark, mentioning that the molecule is fibrous is essential.
It forms supportive tissues in the body ***[1 mark]****.*
Collagen consists of three polypeptide chains ***[1 mark]****.*
These chains form a triple helix ***[1 mark]****.*
The chains are tightly coiled together ***[1 mark]****.*
The tightly coiled chains provide strength to the structure ***[1 mark]****.*
Minerals can bind to the collagen chain ***[1 mark]****.*
This makes it more rigid ***[1 mark]****.*
8 marks are listed, but the mark is given out of 6. This is common in longer exam questions. You would have to be a bit of a mind-reader to hit every mark the examiner thinks of, so to make it fair, there are more mark points than marks. You can only count a maximum of 6, though.

Page 7 — Lipids

1 *Maximum of 8 marks available.*
Triglycerides are made from a glycerol molecule ***[1 mark]*** *and three molecules of fatty acids* ***[1 mark]****. They are formed by condensation reactions* ***[1 mark]****. These reactions result in the formation of ester bonds* ***[1 mark]*** *between the fatty acid and glycerol molecules, with the production of a molecule of water for each fatty acid added* ***[1 mark]****.*
Triglycerides are broken up by hydrolysis reactions ***[1 mark]****, which are the reverse of condensation reactions* ***[1 mark]****, with one molecule of water being added for each fatty acid that's released* ***[1 mark]****.*
It would be possible to get all the marks in this question by using labelled diagrams, as long as all the points listed have been illustrated.

2 *Maximum of 8 marks available.*
A triglyceride consists of glycerol ***[1 mark]*** *and three fatty acid molecules* ***[1 mark]****. A phospholipid has the same basic structure, but one of the fatty acids is replaced by a phosphate group* ***[1 mark]****.*
Triglycerides are hydrophobic / repel water ***[1 mark]****. This is a property of the hydrocarbon chains that are part of the fatty acid molecules* ***[1 mark]****. The phosphate group in a phospholipid is hydrophilic / attracts water* ***[1 mark]****, because it's ionised* ***[1 mark]****. This means that the phospholipid has a hydrophilic 'head' and a hydrophobic 'tail'* ***[1 mark]****.*
The words 'head' and 'tail' are not essential, as long as you have got across the idea that the molecule is partly hydrophobic and partly hydrophilic.

Page 9 — Biochemical Tests for Molecules

1 *Maximum of 14 marks available.*
Test for starch ***[1 mark]*** *by adding iodine in potassium iodide solution* ***[1 mark]****. A positive reaction would have a blue-black colour* ***[1 mark]****.*
Test for reducing sugar ***[1 mark]*** *by heating with Benedict's reagent* ***[1 mark]****. The formation of a brick red precipitate would indicate the presence of reducing sugar* ***[1 mark]****.*
Instead of mentioning red precipitate, you could just describe a colour change from blue to brick red.
Test for non-reducing sugars ***[1 mark]*** *by boiling with hydrochloric acid and then neutralising before doing the Benedict's test* ***[1 mark]****.*
Test for proteins ***[1 mark]*** *by adding sodium hydroxide solution, then copper (II) sulphate solution* ***[1 mark]****. Protein is indicated by a purple colour* ***[1 mark]****.*
Test for lipids ***[1 mark]*** *by adding ethanol then mixing with water* ***[1 mark]****. Lipid is indicated by a milky colour* ***[1 mark]****.*
For full marks, make sure you've mentioned what each biochemical group is, how you test for it and what results you might get.

2 *Maximum of 7 marks available.*
Grind up the leaves in a solvent ***[1 mark]****.*
Add a drop of the extract to strip of chromatography paper ***[1 mark]****.*
Place the end of the paper into a solvent so that the solvent rises up the paper ***[1 mark]****.*
Measure the distance that the solvent has travelled ***[1 mark]*** *and the distance that each spot of pigment has travelled* ***[1 mark]****.*
To identify the pigments calculate their Rf values ***[1 mark]*** *by using the formula Rf = distance travelled by pigment/distance travelled by solvent* ***[1 mark]****.*
Sometimes it is easier or quicker to describe something in a diagram rather than in words (e.g. calculating Rf values). You can get full marks this way as long as the diagram shows what the examiner wants to know.

Answers

Page 11 — Water

1 *Maximum of 12 marks available, from any of the 15 mark points listed.*
Water molecules have two hydrogen atoms and one oxygen atom ***[1 mark]****.*
The hydrogen and oxygen atoms are joined by covalent bonds / sharing electrons ***[1 mark]****.*
Water molecules are polar ***[1 mark]****.*
Polarity leads to the formation of hydrogen bonds between water molecules ***[1 mark]****.*
Water is a solvent ***[1 mark]****.*
Water's polar nature allows water to dissolve polar solutes ***[1 mark]****.*
Water transports substances ***[1 mark]****.*
Substances are transported more easily when dissolved ***[1 mark]****.*
Water has a high specific heat capacity ***[1 mark]****.*
This is due to hydrogen bonds restricting movement ***[1 mark]****.*
This means it's difficult to change the temperature of water ***[1 mark]****.*
This allows cells / tissues to avoid sudden changes in temperature ***[1 mark]****.*
Water has a high latent heat of evaporation ***[1 mark]****.*
This is due to hydrogen bonding ***[1 mark — but only awarded if hydrogen bonding hasn't already been mentioned]****.*
It means the evaporation of water has a considerable cooling effect ***[1 mark]****.*
Be careful to stick to what the question asks for. The fact that ice floats on water and aquatic habitats have stable temperatures aren't things that happen "in living organisms".

Page 15 — Cells, Tissues and Organs, Organelles, Microscopes and Differential Centrifugation

1 *Maximum of 9 marks available, from any of the 4 answers given below:*
Mitochondria [1 mark] *— Large numbers of mitochondria would indicate that the cell used a lot of energy* ***[1 mark]****, because mitochondria are the site of (aerobic) respiration, which releases energy* ***[1 mark]****.*
Chloroplasts [1 mark] *— Large numbers of chloroplasts would be seen in cells that are involved in photosynthesis* ***[1 mark]*** *because the chloroplasts contain chlorophyll, which absorbs light for photosynthesis* ***[1 mark]****.*
Rough endoplasmic reticulum [1 mark] *—*
You find a lot of RER in cells that produce a lot of protein ***[1 mark]*** *because the RER transports protein made in its attached ribosomes* ***[1 mark]***
Ribosomes [1 mark] *— large numbers of ribosomes are found in cells that produce lots of protein* ***[1 mark]*** *because they are the site where proteins are made* ***[1 mark].***
There are 9 marks for this question and 3 organelles have to be mentioned — so it's logical that each organelle provides 3 marks. You get a mark for mentioning the correct organelles, so you need to give 2 pieces of relevant information for each one.

2 a) *Maximum of 2 marks available.*
i) mitochondrion ***[1 mark]***
ii) Golgi apparatus ***[1 mark]***
b) *Maximum of 2 marks available:*
The function of the mitochondrion is to be the site of (aerobic) respiration / provide energy ***[1 mark]****.*
The function of the Golgi apparatus is to package materials made in the cell / to make lysosomes ***[1 mark]****.*
The question doesn't ask you to give the reasons why you identified the organelles as you did, so don't waste time writing your reasons down.

3 *Maximum of 5 marks available.*
The cells are homogenised ***[1 mark]*** *in ice cold isotonic buffer solution* ***[1 mark]****. The cold makes sure that any protein digesting enzymes that are released don't digest the organelles* ***[1 mark]****. The buffer keeps the pH constant* ***[1 mark]*** *and the isotonic solution ensures that osmosis does not occur as this could harm the organelles* ***[1 mark]****.*

Page 17 — Plasma Membranes

1 a) *Maximum of 1 mark available.*
In a triglyceride there are three fatty acids attached to a molecule of glycerol but in a phospholipid one of the fatty acids is replaced by a phosphate group ***[1 mark]****.*
b) *Maximum of 2 marks available.*
Phospholipids are arranged in a double layer / bilayer with fatty acid tails on the inside ***[1 mark]****.*
Fatty acid tails are hydrophobic / non-polar so they prevent the passage of water soluble molecules through the cell membrane ***[1 mark]****.*
Occasionally a question may ask you to show how a single layer of phospholipid molecules would arrange themselves on the surface of a container of water. You should draw the molecules with their hydrophilic phosphate heads in the water and their hydrophobic fatty acid tails sticking up into the air.

Page 21 — Diffusion, Osmosis, Facilitated Diffusion and Active Transport

1 *Maximum of 2 marks available.*
The ions are moving faster / have more kinetic energy ***[1 mark]****, so more ions will diffuse through the membrane in a given time* ***[1 mark]****.*

2 *Maximum of 3 marks available.*
The root hair cell has a more negative / lower water potential than the soil ***[1 mark]*** *because it has a greater concentration of dissolved solutes* ***[1 mark]****.*
So water moves into the cell from the soil by osmosis from a higher water potential to a lower water potential ***[1 mark]****.*

Page 23 — Exchange Surfaces

1 *Maximum of 4 marks available.*
Humans are large multicellular organisms ***[1 mark]****.*
The surface area : volume ratio is small in large organisms ***[1 mark]****, which makes diffusion too slow* ***[1 mark]****.*
Humans need specialised organs with a large enough surface area to keep all their cells supplied with enough oxygen and to remove CO_2 ***[1 mark]****.*

Page 25 — Gas Exchange in Fish and Plants

1 *Maximum of 4 marks available.*
The gills contain many lamellae ***[1 mark]*** *which provide a large surface area for exchange* ***[1 mark]****.*
The counter-current system, where blood and water flow in opposite directions, makes diffusion more efficient ***[1 mark]****.*
Blood is always flowing towards water containing lots of oxygen, which maintains a steep concentration gradient ***[1 mark]****.*

2 *Maximum of 4 marks available.*
It's moist which means gases diffuse faster ***[1 mark]****. There's a large surface area for gas exchange* ***[1 mark]****. Mesophyll cells have thin cell walls and membranes — so there is a short diffusion pathway for gases into the cell* ***[1 mark]****. There are lots of air spaces in the spongy mesophyll layer which allow gases to circulate* ***[1 mark]****.*

Answers

Page 28 — Enzymes

1 *Maximum of 8 marks available, from any of the 9 points below.*
If the solution is too cold, the enzyme will work very slowly ***[1 mark]****.*
This is because, at low temperatures, the molecules move slowly and collisions are less likely between enzyme and substrate molecules ***[1 mark]****.*
The marks above could also be obtained by giving the reverse argument — a higher temperature is best to use because the molecules will move fast enough to give a reasonable chance of collisions.
If the temperature gets too high, the reaction will stop ***[1 mark]****.*
This is because the enzyme is denatured ***[1 mark]*** *— the active site changes and will no longer fit the substrate* ***[1 mark]****.*
Denaturation is caused by increased vibration breaking bonds in the enzyme ***[1 mark]****.*
Enzymes have an optimum pH ***[1 mark]****.*
pH values too far from the optimum cause denaturation ***[1 mark — an explanation of denaturation here will only be awarded a mark if it hasn't been explained earlier]****.*
Denaturation by pH is caused by disruption of ionic bonds, which destabilises the enzyme's tertiary structure ***[1 mark]****.*

2 *Maximum of 4 marks available.*
Chemical X is an enzyme inhibitor ***[1 mark].***
Reason — it reduces an enzyme controlled reaction ***[1 mark].***
The inhibitor is probably competitive ***[1 mark].***
Reason — increasing the concentration of the inhibitor makes it more effective, because if there are a lot of inhibitor molecules they're more likely to reach active sites before the substrate molecules and will block them ***[1 mark]****.*

Page 31 — Digestion and Absorption

1 *Maximum of 4 marks available, from any of the 8 points below.*
Polypeptides are broken down by peptidases ***[1 mark]*** *to form amino acids* ***[1 mark]*** *when peptide bonds are hydrolysed* ***[1 mark]****. Exopeptidases hydrolyse peptide bonds between amino acids on the outside* ***[1 mark]*** *of the polypeptide chain. Endopeptidases hydrolyse peptide bonds between amino acids on the inside* ***[1 mark]*** *of the polypeptide chain. Peptidases are released into the acidic conditions of the stomach in gastric juice* ***[1 mark]*** *and in the duodenum from pancreatic juice* ***[1 mark]*** *and from the epithelial cells lining the small intestine* ***[1 mark]****.*

2 *Maximum of 3 marks available.*
Small, soluble products of digestion (e.g. glucose, amino acids, fatty acids) ***[1 mark]*** *are absorbed into the body through microvilli lining the gut wall* ***[1 mark]****. Absorption is through diffusion, facilitated diffusion and active transport* ***[1 mark if all 3 methods are mentioned]****.*

Section 2 — Genes and Genetic Engineering

Page 33 — Basic Structure of DNA and RNA, Replication of DNA

1 *Maximum of 2 marks available.*
The long length and coiled nature of DNA molecules allows the storage of vast quantities of information ***[1 mark]****.*
You only get the mark here if you've mentioned the length and the coiled nature.
Good at replicating itself because of the two strands being paired ***[1 mark]****.*
The question is worth two marks so you need to mention at least two things.

2 *Maximum of 5 marks available.*
Nucleotides are joined by condensation reactions ***[1 mark]****.*
This happens between the phosphate group and the sugar of the next nucleotide ***[1 mark]****.*
The DNA strands join through hydrogen bonds ***[1 mark]*** *between the base pairs* ***[1 mark]****.*
The final mark is given for at least one accurate diagram showing at least one of the above processes ***[1 mark]****.*
As the question asks for a diagram make sure you do at least one, e.g.:

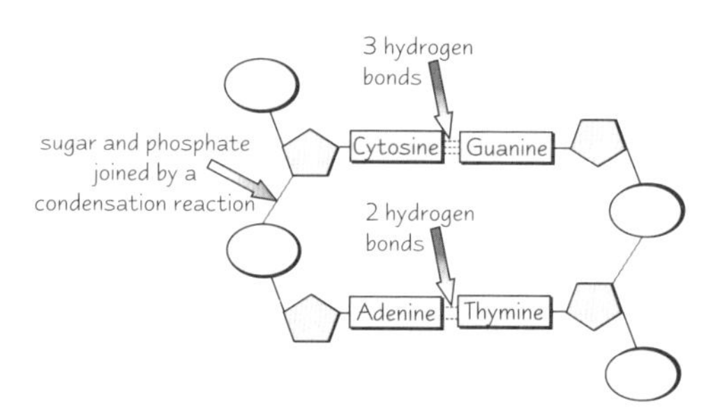

3 *Maximum of 6 marks available.*
DNA strands uncoil and separate ***[1 mark]****.*
Individual free DNA nucleotides pair up with their complementary bases on the template strand ***[1 mark]****.*
DNA polymerase joins the individual nucleotides together ***[1 mark]****.*
Students often forget to mention this enzyme in their answers.
Hydrogen bonds then form between the bases on each strand ***[1 mark].***
Two identical DNA molecules are produced ***[1 mark]****.*
Each of the new molecules contains a single strand from the original DNA molecule and a single new strand ***[1 mark]****.*

Page 35 — Genes and the Genetic Code, Types of RNA

1 *Maximum of 2 marks available.*
Genes are lengths of DNA, found on chromosomes ***[1 mark]****. Genes contain the genetic information that determines the development and characteristics of an organism* ***[1 mark]****.*

2 *Maximum of 2 marks available.*
UUACGUCCGAGA ***[2 marks — 1 mark awarded if one letter is incorrect, no marks for more than one mistake]****.*

Answers

Page 37 — Protein Synthesis, Mutation

1 a) *Maximum of 1 mark available.*
This is a substitution / thymine's been substituted by adenine ***[1 mark]****.*

b) *Maximum of 3 marks available.*
The different base results in the codon / base triplet coding for a different amino acid ***[1 mark]****.*
The new protein formed has a different three-dimensional (tertiary) structure to the original protein ***[1 mark]****.*
This abnormal protein can't function in the same way as the original ***[1 mark]****.*

Page 39 — Mitosis, Cloning

1 a) *Maximum of 6 marks available.*
A = Metaphase ***[1 mark]****, because the chromosomes are lining up at the equator* ***[1 mark]****.*
B = Telophase ***[1 mark]****, because the cytoplasm is dividing to form two new cells* ***[1 mark]****.*
C = Anaphase ***[1 mark]****, because the centromeres have divided and the chromatids are moving to opposite poles* ***[1 mark]****.*
If you've learned the diagrams of what happens at each stage of mitosis, this should be a breeze. That's why it'd be a total disaster if you lost three marks for forgetting to give reasons for your answers. Always read the question properly and do exactly what it tells you to do.

b) *Maximum of 3 marks available:*
X = Chromatid ***[1 mark]****.*
Y = Centromere ***[1 mark]****.*
Z = Spindle fibre ***[1 mark]****.*

2 a) *Maximum of 1 mark available.*
Interphase ***[1 mark]****.*

b) *Maximum of 1 mark available.*
Prophase ***[1 mark]****.*

Page 41 — Gametes, Meiosis and Sexual Reproduction

1 a) *Maximum of 3 marks available.*
A = 46 ***[1 mark]****.*
B = 23 ***[1 mark]****.*
C = 23 ***[1 mark]****.*

b) *Maximum of 2 marks available, from any of the points below.*
Normal body cells have two copies of each chromosome, which they inherit from their parents ***[1 mark]****.*
Gametes have to have half the number of chromosomes so that when fertilisation takes place, the resulting embryo will have the correct diploid number ***[1 mark]****.*
If the gametes had a diploid number, the resulting offspring would have twice the number of chromosomes that it should have ***[1 mark]****.*

Page 43 — Genetic Engineering

1 a) *Maximum of 2 marks available.*
Extract all the DNA from the cell ***[1 mark]****.*
Identify the gene and use a restriction endonuclease enzyme to cut it out ***[1 mark]****.*

b) *Maximum of 2 marks available.*
Add the same restriction endonuclease enzyme to bacterial plasmids (the bacterial DNA) ***[1 mark]****.*
Insert the gene into the treated plasmids and add ligase to attach the gene to bacterial DNA ***[1 mark]****.*
Don't panic if the question mentions organisms you haven't learnt about. If you read it carefully it will contain familiar ideas. This is the examiners' way of seeing whether you understand the main ideas and if you can apply what you've learnt.

2 a) *Restriction endonuclease* ***[1 mark]****.*

b) *Ligase* ***[1 mark]****.*

Page 45 — Genetic Engineering

1 a) *Maximum of 1 mark available.*
Replica plating ***[1 mark]****.*

b) *Maximum of 5 marks available.*
Bacteria that have been transformed / have taken up the plasmid contain both the human gene and the gene for antibiotic resistance ***[1 mark]****.*
After being mixed with plasmids, the bacteria are cultured on an agar plate called the master plate ***[1 mark]****.*
Then a sterile velvet pad is pressed onto the master plate, which picks up some bacteria from each colony ***[1 mark]****.*
The pad is pressed onto a fresh agar plate, containing an antibiotic, and some of the bacteria from each colony are transferred onto the agar surface ***[1 mark]****.*
Only transformed bacteria can grow and reproduce on the replica plate — the others don't contain the antibiotic-resistant gene, so they stop growing ***[1 mark]****.*

Page 47 — Gene Therapy

1 a) *Maximum of 4 marks available.*
The CFTR protein is a channel protein which allows chloride ions to move out of the cell ***[1 mark]****. When there's a mutation in the CFTR gene, the CFTR protein is missing an amino acid which affects its 3D shape* ***[1 mark]****.*
It can't let chloride ions through, so they build up in the cell ***[1 mark]****.*
The build up of ions means that less water moves out of the cell by osmosis ***[1 mark]****.*

b) *Maximum of 5 marks available.*
EITHER — Using liposomes as vectors:
Recombinant DNA technology is used to put healthy CFTR genes into plasmids ***[1 mark]****.*
The plasmids are wrapped in lipid droplets to form liposomes ***[1 mark]****.*
These are sprayed into the airways using an aerosol ***[1 mark]****.*
The liposomes fuse with the cell membranes of the airway epithelial cells, releasing the plasmids into the cells ***[1 mark]****.*
The human DNA is expressed and normal CFTR is produced ***[1 mark]****.*
You don't need to go into details about recombinant DNA technology here — the point of the question is how the vector is used.
OR — Using viruses as vectors:
Adenoviruses are treated to make the viral DNA harmless ***[1 mark]****.*
The inactivated viral DNA and the healthy CFTR gene are joined to form recombinant DNA ***[1 mark]****.*
The modified viruses are sprayed into the lungs ***[1 mark]****.*
The viruses inject the airway epithelial cells with their DNA, which also injects the healthy CFTR gene ***[1 mark]****.*
The CFTR gene is expressed by the epithelial cells in the usual way ***[1 mark]****.*

2 a) *Maximum of 3 marks available.*
The human gene coding for production of factor VIII is cloned and combined with a promoter sequence of DNA ***[1 mark]****.*
Mature eggs are taken from the sheep's ovary, fertilised in vitro and then have the factor VIII gene micro-injected into them ***[1 mark]****.*
The zygotes are cultured in vitro and are implanted into sheep surrogate mothers ***[1 mark]****.*
Don't be put off just because it's a different protein from the one described on page 47 — the process is exactly the same.

b) *Maximum of 1 mark available, from any of the following:*
Less risk of transfer of infectious agents from donor ***[1 mark]****.*
More protein produced from easily accessible source ***[1 mark]****.*
Less risk of allergic response from other blood proteins ***[1 mark]****.*

Answers

Page 49 — Genetic Fingerprinting

1 a) *Maximum of 1 mark available.*
Genetic / DNA fingerprinting ***[1 mark]****.*
b) *Maximum of 5 marks available.*
A blood/semen/sweat/hair/skin sample would be taken from the suspect ***[1 mark]****.*
The DNA from the sample would be cut into fragments by the specific restriction endonucleases ***[1 mark]****.*
The DNA fragments from both samples would be separated out by size using electrophoresis ***[1 mark]****.*
Radioactive gene probes would be used to make the DNA fragment patterns of both samples visible ***[1 mark]****.*
The bands on the photographic film would be compared to those of DNA samples found at the scene of the crime ***[1 mark]****.*
'Explain' in an exam question tests your ability to apply your knowledge. You must give your answer in the context of the question. Simply describing the technique of genetic fingerprinting wouldn't get you full marks

Section 3 — Physiology and Transport

Page 51 — The Mammalian Heart

1 *Maximum of 3 marks available.*
Pressure increases in atria during atrial systole and in ventricles during ventricular systole ***[1 mark]****.*
Pressure decreases in atria during atrial diastole and in the ventricles during ventricular diastole ***[1 mark]****.*
There is always more pressure on the left side of the heart due to extra muscle tissue producing more force ***[1 mark]****.*
This question doesn't ask you to describe the cardiac cycle — it specifically asks you to describe the pressure changes in diastole and systole. Make sure you mention both atria and ventricles in your answer.

2 *Maximum of 6 marks available.*
The valves only open one way ***[1 mark]****.*
Whether they open or close depends on the relative pressure of the heart chambers ***[1 mark]****.*
If the pressure is greater behind a valve (i.e. there's blood in the chamber behind it) ***[1 mark]****, it's forced open, to let the blood travel in the right direction* ***[1 mark]****.*
When the blood goes through the valve, the pressure is greater above the valve ***[1 mark]****, which forces it shut, preventing blood from flowing back into the chamber* ***[1 mark]****.*
Here you need to explain how valves function in relation to blood flow, rather than just in relation to relative pressures.

Page 53 — Transport Systems

1 *Maximum of 4 marks available.*
Alveoli ***[1 mark]*** *and villi* ***[1 mark]****.*
Both are small so have high surface area : volume ratio ***[1 mark]****, both have thin membranes and are close to capillaries so there's a short diffusion pathway for exchange of substances* ***[1 mark]****.*
Surface area to volume ratio again. It comes up everywhere — if in doubt, just put surface area to volume ratio. Best if you know what you're talking about though, so make sure you really understand the idea behind those magic words. Have another look at page 22 if you haven't quite got it yet.

2 *Maximum of 6 marks available.*
Diffusion alone wouldn't be fast enough to carry out enough exchange of materials ***[1 mark]****. This is because larger organisms tend to have a tough outer surface, so it's harder for substances to enter that way* ***[1 mark]****. Once the substances were inside they would need to travel too far to reach the cells deepest inside the body in time* ***[1 mark]****. Larger organisms have a low surface area to volume ratio* ***[1 mark]*** *and need a large quantity of substances* ***[1 mark]*** *(multicellular animals have greater demand due to a higher metabolic rate)* ***[1 mark]****.*

Page 55 — Blood, Tissue Fluid and Lymph

1 *Maximum of 5 marks available.*
Tissue fluid moves out of capillaries due to the pressure gradient ***[1 mark].***
At the arteriole end, pressure in capillary beds is greater than pressure in tissue fluid outside capillaries ***[1 mark]****.*
This means fluid from blood is forced out of the capillaries ***[1 mark]****.*
Fluid loss means the water potential of blood capillaries is lower than that of tissue fluid ***[1 mark]****.*
So fluid moves into the capillaries at the vein end by osmosis ***[1 mark]****.*

2 *Maximum of 3 marks available.*
They have no nucleus to leave more room for haemoglobin ***[1 mark]****.*
Their biconcave disc shape means they have a large surface area for oxygen diffusion ***[1 mark]****.*
They can squeeze through small capillaries because they have an elastic membrane ***[1 mark]****.*
Exam questions relating to structure and function are a favourite. Find a way (cartoons, poems, tables) to remember the name, structure and function all together to avoid confusion.

Page 57 — Haemoglobin and Oxygen Transport

1 *Maximum of 3 marks available.*
The first oxygen molecule finds it hard to join onto the haemoglobin so the graph is not very steep. ***[1 mark]****.*
This oxygen molecule changes the shape of the haemoglobin molecule making it easier for other molecules to join on and so the graph becomes steeper ***[1 mark]****.*
As the molecule becomes more saturated it becomes harder for oxygen to attach and so the graph becomes less steep ***[1 mark]****.*

Page 59 — Control of Heartbeat

1 *Maximum of 4 marks available.*
The low blood pressure would be detected by pressure receptors in the arteries ***[1 mark]****.*
The pressure receptors would send impulses to the cardiovascular centre in the medulla in the brain ***[1 mark]****.*
The cardiovascular centre would then send impulses to the SAN to speed up the heart rate ***[1 mark]****.*
This would cause blood pressure to increase ***[1 mark]****.*
Make sure you've said that speeding up the heart rate is done to increase blood pressure. It might seem a bit obvious, but you should still write it to show you understand the whole process.

2 *Maximum of 6 marks available.*
The SAN produces a stimulus ***[1 mark]*** *which causes the left and right atrial muscles to contract* ***[1 mark]****.*
The AVN picks up the stimulus ***[1 mark]*** *and sends impulses through the Bundle of His* ***[1 mark]*** *and on to the Purkyne tissue in the ventricle walls* ***[1 mark]****, which enable the left and right ventricles to contract* ***[1 mark]****.*
Examiners hate the use of 'messages' or 'signals' for nerve impulses — use the correct terms so you don't throw away easy marks.

Answers

Page 61 — Ventilation and Types of Respiration

1 *Maximum of 4 marks available.*
An increase of carbon dioxide causes a decrease in blood pH ***[1 mark]****.*
Chemoreceptors detect this change and send a signal to the medulla ***[1 mark]****.*
The medulla increases the number of nerve impulses to the intercostal muscles and the diaphragm ***[1 mark]****.*
The rate and depth of breathing increases and the amount of gaseous exchange in the lungs and at the respiring tissues increases ***[1 mark]****.*

2 *Maximum of 3 marks available.*
It is transported to the liver in the blood ***[1 mark]****.*
It is converted to pyruvate and then to glucose using oxygen ***[1 mark]****.*
The glucose is either taken back to respiring cells for energy or is stored as glycogen ***[1 mark]****.*

Page 65 — Roots, Xylem and Phloem, Transpiration

1 *Maximum of 6 marks available.*
a) Water can pass through the cell walls en route to the xylem ***[1 mark]****.*
This is the apoplast pathway ***[1 mark]****.*
b) Each endodermis cell has a waxy Casparian strip in its cell wall that water can't penetrate ***[1 mark]****.*
This blocks the apoplast pathway and allows selective absorption through the cell membrane ***[1 mark]****.*
c) Plasmodesmata connect the cytoplasm of adjacent cells so water can travel from cell to cell through them ***[1 mark]****.*
This is the symplast pathway ***[1 mark]****.*

2 *Maximum of 10 marks available.*
Distribution can be explained in words or by diagrams, whichever you find easier. In either case, these are the key points:
In the stem —
Xylem and phloem towards the outside, with the phloem outside the xylem ***[1 mark]****.*
In the root —
The xylem and phloem in the centre, with the phloem outside the xylem ***[1 mark]****.*
In the leaf —
Main vein in the midrib, smaller veins in the rest of the leaf, with xylem above the phloem ***[1 mark]****.*
Function of xylem — to transport water and minerals ***[1 mark]*** *and to provide support* ***[1 mark]****.*
Stem is subjected to bending forces ***[1 mark]****.*
This is best resisted by strengthening around the outside ***[1 mark]****.*
Root is subjected to crushing forces ***[1 mark]****.*
This is best resisted by strengthening in the centre ***[1 mark]****.*
The leaf is thin and needs support throughout the tissue ***[1 mark]****.*

3 *Maximum of 8 marks available.*
Up to 4 marks for correctly naming each factor. Up to 4 marks for explaining the effect of each one.
Light ***[1 mark]****.*
The stomata close in the dark, so transpiration cannot happen ***[1 mark]****.*
Temperature ***[1 mark]****.*
Increasing temperature speeds up the movement of the water molecules ***[1 mark]****.*
Humidity ***[1 mark]****.*
Increasing humidity reduces the diffusion gradient between the leaf and the air, so transpiration slows down (or reverse argument for decreasing humidity) ***[1 mark]****.*
Air movement / wind ***[1 mark]****.*
Dispersion of the water molecules in the air around the stomata maintains the diffusion gradient ***[1 mark]****.*

Page 67 — Translocation

1 *Maximum of 2 marks available.*
Greatest radioactivity would be seen in growing areas of the plant (e.g. young leaves, the growing point of the stem and the growing points in the roots) ***[1 mark]****. These areas would receive most sugar because they are where most energy is needed for growth* ***[1 mark]****.*

2 *Maximum of 4 marks available.*
Sugars are actively transported into the sieve tubes at the source end ***[1 mark]****.*
This decreases the water potential of the sieve tubes ***[1 mark]****.*
This causes water to flow in by osmosis ***[1 mark]****.*
This means pressure is increased inside the sieve tubes at the source end ***[1 mark]****.*
I think this is a pretty nasty question. If you got it all right first time you're probably a genius. If you didn't, you're probably not totally clear yet about the pressure gradient idea. It's just that when a cell fills with fluid, the molecules inside are under more and more pressure. If there's a high concentration of sugar in a cell, this draws in water by osmosis, and so increases the pressure inside the cell.

Index

Index

Index

Index

Index